The Hidden Life of Trees

foreword by TIM FLANNERY

PETER WOHLLEBEN

TRANSLATION BY JANE BILLINGHURST

The Hidden Life of TREES

What They Feel, How They Communicate

Discoveries from a Secret World

WILLIAM COLLINS

William Collins
An imprint of HarperCollins*Publishers*
1 London Bridge Street
London SE1 9GF

WilliamCollinsBooks.com

HarperCollins*Publishers*
1st Floor, Watermarque Building, Ringsend Road
Dublin 4, Ireland

This paperback edition first published in
the United Kingdom by William Collins in 2017

22 21
16

Originally published in Germany as *Das geheime Leben der Bäume. Was sie fühlen, wie sie kommunizieren. Die Entdeckung einer verborgenen Welt*. By Peter Wohlleben © 2015 by Ludwig Verlag, a division of Verlagsgruppe Random House GmbH, München, Germany.

The Hidden Life of Trees first published in the English language in hardback by Greystone Books Ltd in 2016, Vancouver, Canada.

A catalogue record for this book is
available from the British Library

Copyediting by Shirarose Wilensky
Jacket and interior design by Nayeli Jimenez
Jacket and interior illustrations by Briana Garelli

ISBN 978-0-00-821843-0

Printed and bound in Great Britain by
CPI Group (UK) Ltd, Croydon

MIX
Paper from
responsible sources
FSC™ C007454

This book is produced from independently certified FSC paper
to ensure responsible forest management

For more information visit: www.harpercollins.co.uk/green

TABLE OF CONTENTS

Foreword by Tim Flannery vii
Introduction to the English Edition x
Introduction xiii

1 / Friendships *1*
2 / The Language of Trees *6*
3 / Social Security *14*
4 / Love *19*
5 / The Tree Lottery *25*
6 / Slowly Does It *31*
7 / Forest Etiquette *37*
8 / Tree School *43*
9 / United We Stand, Divided We Fall *49*
10 / The Mysteries of Moving Water *56*
11 / Trees Aging Gracefully *60*
12 / Mighty Oak or Mighty Wimp? *68*
13 / Specialists *73*
14 / Tree or Not Tree? *79*
15 / In the Realm of Darkness *85*
16 / Carbon Dioxide Vacuums *93*

17 / Woody Climate Control *99*
18 / The Forest as Water Pump *105*
19 / Yours or Mine? *113*
20 / Community Housing Projects *125*
21 / Mother Ships of Biodiversity *131*
22 / Hibernation *136*
23 / A Sense of Time *147*
24 / A Question of Character *151*
25 / The Sick Tree *155*
26 / Let There Be Light *162*
27 / Street Kids *169*
28 / Burnout *180*
29 / Destination North! *186*
30 / Tough Customers *195*
31 / Turbulent Times *201*
32 / Immigrants *211*
33 / Healthy Forest Air *221*
34 / Why Is the Forest Green? *227*
35 / Set Free *233*
36 / More Than Just a Commodity *241*

Note from a Forest Scientist by Dr. Suzanne Simard *247*
Acknowledgments *251*
Notes *252*
Index *261*

FOREWORD

WE READ IN fairy tales of trees with human faces, trees that can talk, and sometimes walk. This enchanted forest is the kind of place, I feel sure, that Peter Wohlleben inhabits. His deep understanding of the lives of trees, reached through decades of careful observation and study, reveals a world so astonishing that if you read his book, I believe that forests will become magical places for you, too.

One reason that many of us fail to understand trees is that they live on a different time scale than us. One of the oldest trees on Earth, a spruce in Sweden, is more than 9,500 years old. That's 115 times longer than the average human lifetime. Creatures with such a luxury of time on their hands can afford to take things at a leisurely pace. The electrical impulses that pass through the roots of trees, for example, move at the slow rate of one third of an inch per second. But why, you might ask, do trees pass electrical impulses through their tissues at all?

The answer is that trees need to communicate, and electrical impulses are just one of their many means of communication. Trees also use the senses of smell and taste for communication. If a giraffe starts eating an African acacia, the tree releases a chemical into the air that signals that a threat is at hand. As the chemical drifts through the air and reaches other trees, they "smell" it and are warned of the danger. Even before the giraffe reaches them, they begin producing toxic chemicals. Insect pests are dealt with slightly differently. The saliva of leaf-eating insects can be "tasted" by the leaf being eaten. In response, the tree sends out a chemical signal that attracts predators that feed on that particular leaf-eating insect. Life in the slow lane is clearly not always dull.

But the most astonishing thing about trees is how social they are. The trees in a forest care for each other, sometimes even going so far as to nourish the stump of a felled tree for centuries after it was cut down by feeding it sugars and other nutrients, and so keeping it alive. Only some stumps are thus nourished. Perhaps they are the parents of the trees that make up the forest of today. A tree's most important means of staying connected to other trees is a "wood wide web" of soil fungi that connects vegetation in an intimate network that allows the sharing of an enormous amount of information and goods. Scientific research aimed at understanding the astonishing abilities of this partnership between fungi and plant has only just begun.

The reason trees share food and communicate is that they need each other. It takes a forest to create a microclimate suitable for tree growth and sustenance. So it's not surprising that isolated trees have far shorter lives than those living connected together in forests. Perhaps the saddest plants of all are those we have enslaved in our agricultural systems. They seem to have lost the ability to communicate, and, as Wohlleben says, are isolated by their silence. "Perhaps farmers can learn from the forests and breed a little more wildness back into their grain and potatoes," he advocates, "so that they'll be more talkative in the future."

Opening this book, you are about to enter a wonderland. Enjoy it.

TIM FLANNERY

INTRODUCTION TO THE ENGLISH EDITION

WHEN I WROTE this book, I wanted to describe my experiences in the forest I manage in the Eifel mountains in Germany and record what the trees had taught me. As soon as the German edition of the book was published, it was clear that the story I had to tell struck a chord with many, many people. My message, though grounded in a forest I interact with almost every day, is a message that applies to forests and woodlands around the world.

I am most familiar with the struggles and strategies of beeches and oaks, and with the contrast between deciduous forests that plan their own futures and coniferous forests planted for commercial gain. However, the struggles and strategies in forests left to their own devices, and the tension created when forests are planted instead of evolving at their own pace, are issues that resonate far beyond my experiences in Hümmel.

I encourage you to look around where you live. What dramas are being played out in wooded areas you can explore?

How are commerce and survival balanced in the forests and woodlands you know? This book is a lens to help you take a closer look at what you might have taken for granted. Slow down, breathe deep, and look around. What can you hear? What do you see? How do you feel?

My story also explains why forests matter on a global scale. Trees are important, but when trees unite to create a fully functioning forest, you really can say that the whole is greater than its parts. Your trees may not function exactly as my trees do, and your forest might look a little different, but the underlying narrative is the same: forests matter at a more fundamental level than most of us realize.

Before you plunge into this book to find out what I have discovered just by stepping outside my back door, I want to tell you a story about Yellowstone National Park in the United States to show just how vital undisturbed forests and woodlands are to the future of our planet and how our appreciation for trees affects the way we interact with the world around us.

It all starts with the wolves. Wolves disappeared from Yellowstone, the world's first national park, in the 1920s. When they left, the entire ecosystem changed. Elk herds in the park increased their numbers and began to make quite a meal of the aspens, willows, and cottonwoods that lined the streams. Vegetation declined and animals that depended on the trees left. The wolves were absent for seventy years. When they returned, the elks' languorous browsing days were over. As the wolf packs kept the herds on the move, browsing diminished, and the trees sprang back. The roots of cottonwoods and willows once again stabilized stream banks and slowed the flow of water. This, in turn, created space for animals such as beavers to return. These industrious builders could now find the materials they needed to construct their lodges and raise their families. The animals that depended on the riparian meadows came back, as well. The wolves turned out to be better stewards of the land than people, creating conditions that allowed the trees to grow and exert their influence on the landscape.

My hope is that the wolves' stewardship of natural processes in Yellowstone will help people appreciate the complex ways that trees interact with their environment, how our interactions with forests affect their success, and the role forests play in making our world the kind of place where we want to live. Apart from that, forests hide wonders that we are only just beginning to explore. I invite you to enter my world.

INTRODUCTION

WHEN I BEGAN my professional career as a forester, I knew about as much about the hidden life of trees as a butcher knows about the emotional life of animals. The modern forestry industry produces lumber. That is to say, it fells trees and then plants new seedlings. If you read the professional literature, you quickly get the impression that the well-being of the forest is only of interest insofar as it is necessary for optimizing the lumber industry. That is enough for what foresters do day to day, and eventually it distorts the way they look at trees. Because it was my job to look at hundreds of trees every day—spruce, beeches, oaks, and pines—to assess their suitability for the lumber mill and their market value, my appreciation of trees was also restricted to this narrow point of view.

About twenty years ago, I began to organize survival training and log-cabin tours for tourists. Then I added a place in the forest where people can be buried as an alternative to traditional graveyards, and an ancient forest preserve. In conversations with the many visitors who came, my view of the forest changed once again. Visitors were enchanted by crooked, gnarled trees I would previously have dismissed because of their low commercial value. Walking with my visitors, I learned to pay attention to more than just the quality of the trees' trunks. I began to notice bizarre root shapes, peculiar growth patterns, and mossy cushions on bark. My love of Nature—something I've had since I was six years old—was reignited. Suddenly, I was aware of countless wonders I could hardly explain even to myself. At the same time, Aachen University (RWTH Aachen) began conducting regular scientific research programs in the forest I manage. During the course of this research, many questions were answered, but many more emerged.

Life as a forester became exciting once again. Every day in the forest was a day of discovery. This led me to unusual ways of managing the forest. When you know that trees experience pain and have memories and that tree parents live together with their children, then you can no longer just chop them down and disrupt their lives with large machines. Machines have been banned from the forest for a couple of decades now, and if a few individual trees need to be harvested from time to time, the work is done with care by foresters using horses instead. A healthier—perhaps you could even say happier—forest is considerably more productive, and that means it is also more profitable.

This argument convinced my employer, the community of Hümmel, and now this tiny village in the Eifel mountains will not consider any other way of managing their forest. The trees are breathing a collective sigh of relief and revealing even more of their secrets, especially those stands growing in the newly established preserves, where they are left completely undisturbed. I will never stop learning from them, but even what I have learned so far under their leafy canopy exceeds anything I could ever have dreamed of.

I invite you to share with me the joy trees can bring us. And, who knows, perhaps on your next walk in the forest, you will discover for yourself wonders great and small.

BIRCH

1

— FRIENDSHIPS —

YEARS AGO, I stumbled across a patch of strange-looking mossy stones in one of the preserves of old beech trees that grows in the forest I manage. Casting my mind back, I realized I had passed by them many times before without paying them any heed. But that day, I stopped and bent down to take a good look. The stones were an unusual shape: they were gently curved with hollowed-out areas. Carefully, I lifted the moss on one of the stones. What I found underneath was tree bark. So, these were not stones, after all, but old wood. I was surprised at how hard the "stone" was, because it usually takes only a few years for beechwood lying on damp ground to decompose. But what surprised me most was that I couldn't lift the wood. It was obviously attached to the ground in some way.

I took out my pocketknife and carefully scraped away some of the bark until I got down to a greenish layer. Green?

This color is found only in chlorophyll, which makes new leaves green; reserves of chlorophyll are also stored in the trunks of living trees. That could mean only one thing: this piece of wood was still alive! I suddenly noticed that the remaining "stones" formed a distinct pattern: they were arranged in a circle with a diameter of about 5 feet. What I had stumbled upon were the gnarled remains of an enormous ancient tree stump. All that was left were vestiges of the outermost edge. The interior had completely rotted into humus long ago—a clear indication that the tree must have been felled at least four or five hundred years earlier. But how could the remains have clung onto life for so long?

Living cells must have food in the form of sugar, they must breathe, and they must grow, at least a little. But without leaves—and therefore without photosynthesis—that's impossible. No being on our planet can maintain a centuries-long fast, not even the remains of a tree, and certainly not a stump that has had to survive on its own. It was clear that something else was happening with this stump. It must be getting assistance from neighboring trees, specifically from their roots. Scientists investigating similar situations have discovered that assistance may either be delivered remotely by fungal networks around the root tips—which facilitate nutrient exchange between trees[1]—or the roots themselves may be interconnected.[2] In the case of the stump I had stumbled upon, I couldn't find out what was going on, because I didn't want to injure the old stump by digging around it, but one thing was clear: the surrounding beeches were pumping sugar to the stump to keep it alive.

If you look at roadside embankments, you might be able to see how trees connect with each other through their root systems. On these slopes, rain often washes away the soil, leaving the underground networks exposed. Scientists in the Harz mountains in Germany have discovered that this really is a case of interdependence, and most individual trees of the same species growing in the same stand are connected to each other through their root systems. It appears that nutrient exchange and helping neighbors in times of need is the rule, and this leads to the conclusion that forests are superorganisms with interconnections much like ant colonies.

Of course, it makes sense to ask whether tree roots are simply wandering around aimlessly underground and connecting up when they happen to bump into roots of their own kind. Once connected, they have no choice but to exchange nutrients. They create what looks like a social network, but what they are experiencing is nothing more than a purely accidental give and take. In this scenario, chance encounters replace the more emotionally charged image of active support, though even chance encounters offer benefits for the forest ecosystem. But Nature is more complicated than that. According to Massimo Maffei from the University of Turin, plants—and that includes trees—are perfectly capable of distinguishing their own roots from the roots of other species and even from the roots of related individuals.[3]

But why are trees such social beings? Why do they share food with their own species and sometimes even go so far as to nourish their competitors? The reasons are the same as for human communities: there are advantages to working together. A tree is not a forest. On its own, a tree cannot establish a consistent local climate. It is at the mercy of wind and weather. But together, many trees create an ecosystem that moderates extremes of heat and cold, stores a great deal of water, and generates a great deal of humidity. And in this protected environment, trees can live to be very old. To get to this point, the community must remain intact no matter what. If every tree were looking out only for itself, then quite a few of them would never reach old age. Regular fatalities would result in many large gaps in the tree canopy, which would make it easier for storms to get inside the forest and uproot more trees. The heat of summer would reach the forest floor and dry it out. Every tree would suffer.

Every tree, therefore, is valuable to the community and worth keeping around for as long as possible. And that is why even sick individuals are supported and nourished until they recover. Next time, perhaps it will be the other way round, and the supporting tree might be the one in need of assistance. When thick silver-gray beeches behave like this, they remind me of a herd of elephants. Like the herd, they, too, look after their own, and they help their sick and weak back up onto their feet. They are even reluctant to abandon their dead.

Every tree is a member of this community, but there are different levels of membership. For example, most stumps rot away into humus and disappear within a couple of hundred years (which is not very long for a tree). Only a few individuals are kept alive over the centuries, like the mossy "stones" I've just described. What's the difference? Do tree societies have second-class citizens just like human societies? It seems they do, though the idea of "class" doesn't quite fit. It is rather the degree of connection—or maybe even affection—that decides how helpful a tree's colleagues will be.

You can check this out for yourself simply by looking up into the forest canopy. The average tree grows its branches out until it encounters the branch tips of a neighboring tree of the same height. It doesn't grow any wider because the air and better light in this space are already taken. However, it heavily reinforces the branches it has extended, so you get the impression that there's quite a shoving match going on up there. But a pair of true friends is careful right from the outset not to grow overly thick branches in each other's direction. The trees don't want to take anything away from each other, and so they develop sturdy branches only at the outer edges of their crowns, that is to say, only in the direction of "non-friends." Such partners are often so tightly connected at the roots that sometimes they even die together.

As a rule, friendships that extend to looking after stumps can only be established in undisturbed forests. It could well be that all trees do this and not just beeches. I myself have observed oak, fir, spruce, and Douglas fir stumps that were still alive long after the trees had been cut down. Planted forests, which is what most of the coniferous forests in Central Europe are, behave more like the street kids I describe in chapter 27. Because their roots are irreparably damaged when they are planted, they seem almost incapable of networking with one another. As a rule, trees in planted forests like these behave like loners and suffer from their isolation. Most of them never have the opportunity to grow old anyway. Depending on the species, these trees are considered ready to harvest when they are only about a hundred years old.

2

— THE LANGUAGE OF TREES —

According to the dictionary definition, language is what people use when we talk to each other. Looked at this way, we are the only beings who can use language, because the concept is limited to our species. But wouldn't it be interesting to know whether trees can also talk to each other? But how? They definitely don't produce sounds, so there's nothing we can hear. Branches creak as they rub against one another and leaves rustle, but these sounds are caused by the wind and the tree has no control over them. Trees, it turns out, have a completely different way of communicating: they use scent.

Scent as a means of communication? The concept is not totally unfamiliar to us. Why else would we use deodorants and perfumes? And even when we're not using these products, our own smell says something to other people, both consciously and subconsciously. There are some people who seem to have no smell at all; we are strongly attracted to others because of their aroma. Scientists believe pheromones in sweat are a decisive factor when we choose our partners—in other words, those with whom we wish to procreate. So it seems fair to say that we possess a secret language of scent, and trees have demonstrated that they do as well.

For example, four decades ago, scientists noticed something on the African savannah. The giraffes there were feeding on umbrella thorn acacias, and the trees didn't like this one bit. It took the acacias mere minutes to start pumping toxic substances into their leaves to rid themselves of the large herbivores. The giraffes got the message and moved on to other trees in the vicinity. But did they move on to trees close by? No, for the time being, they walked right by a few trees and resumed their meal only when they had moved about 100 yards away.

The reason for this behavior is astonishing. The acacia trees that were being eaten gave off a warning gas (specifically, ethylene) that signaled to neighboring trees of the same species that a crisis was at hand. Right away, all the forewarned trees also pumped toxins into their leaves to prepare themselves. The giraffes were wise to this game and therefore moved farther away to a part of the savannah where they could find trees that were oblivious to what was going on. Or else they moved upwind. For the scent messages are carried to nearby trees on the breeze, and if the animals walked upwind, they could find acacias close by that had no idea the giraffes were there.

Similar processes are at work in our forests here at home. Beeches, spruce, and oaks all register pain as soon as some creature starts nibbling on them. When a caterpillar takes a hearty bite out of a leaf, the tissue around the site of the damage changes. In addition, the leaf tissue sends out electrical signals, just as human tissue does when it is hurt. However, the signal is not transmitted in milliseconds, as human signals are; instead, the plant signal travels at the slow speed of a third of an inch per minute.[4] Accordingly, it takes an hour or so before defensive compounds reach the leaves to spoil the pest's meal. Trees live their lives in the really slow lane, even when they are in danger. But this slow tempo doesn't mean that a tree is not on top of what is happening in different parts of its structure. If the roots find themselves in trouble, this information is broadcast throughout the tree, which can trigger the leaves to release scent compounds. And not just any old scent compounds, but compounds that are specifically formulated for the task at hand.

This ability to produce different compounds is another feature that helps trees fend off attack for a while. When it comes to some species of insects, trees can accurately identify which bad guys they are up against. The saliva of each species is different, and trees can match the saliva to the insect. Indeed, the match can be so precise that trees can release pheromones that summon specific beneficial predators. The beneficial predators help trees by eagerly devouring the insects that are bothering them. For example, elms and pines call on small parasitic wasps that lay their eggs inside leaf-eating caterpillars.[5] As the wasp larvae develop, they devour the larger caterpillars bit by bit from the inside out. Not a nice way to die. The result, however, is that the trees are saved from bothersome pests and can keep growing with no further damage.

The fact trees can recognize saliva is, incidentally, evidence for yet another skill they must have. For if they can identify saliva, they must also have a sense of taste.

A drawback of scent compounds is that they disperse quickly in the air. Often they can be detected only within a range of about 100 yards. Quick dispersal, however, also has advantages. As the transmission of signals inside the tree is very slow, a tree can cover long distances much more quickly through the air if it wants to warn distant parts of its own structure that danger lurks. A specialized distress call is not always necessary when a tree needs to mount a defense against insects. The animal world simply registers the tree's basic chemical alarm call. It then knows some kind of attack is taking place and predatory species should mobilize. Whoever is hungry for the kinds of critters that attack trees just can't stay away.

Trees can also mount their own defense. Oaks, for example, carry bitter, toxic tannins in their bark and leaves. These either kill chewing insects outright or at least affect the leaves' taste to such an extent that instead of being deliciously crunchy, they become biliously bitter. Willows produce the defensive compound salicylic acid, which works in much the same way. But not on us. Salicylic acid is a precursor of aspirin, and tea made from willow bark can relieve headaches and bring down fevers. Such defense mechanisms, of course, take time. Therefore, a combined approach is crucially important for arboreal early-warning systems.

Trees don't rely exclusively on dispersal in the air, for if they did, some neighbors would not get wind of the danger. Dr. Suzanne Simard of the University of British Columbia in Vancouver has discovered that they also warn each other using chemical signals sent through the fungal networks around their root tips, which operate no matter what the weather.[6] Surprisingly, news bulletins are sent via the roots not only by means of chemical compounds but also by means of electrical impulses that travel at the speed of a third of an inch per second. In comparison with our bodies, it is, admittedly, extremely slow. However, there are species in the animal kingdom, such as jellyfish and worms, whose nervous systems conduct impulses at a similar speed.[7] Once the latest news has been broadcast, all oaks in the area promptly pump tannins through their veins.

Tree roots extend a long way, more than twice the spread of the crown. So the root systems of neighboring trees inevitably intersect and grow into one another—though there are always some exceptions. Even in a forest, there are loners, would-be hermits who want little to do with others. Can such antisocial trees block alarm calls simply by not participating? Luckily, they can't. For usually there are fungi present that act as intermediaries to guarantee quick dissemination of news. These fungi operate like fiber-optic Internet cables. Their thin filaments penetrate the ground, weaving through it in almost unbelievable density. One teaspoon of forest soil contains many miles of these "hyphae."[8] Over centuries, a single fungus can cover many square miles and network an entire forest. The fungal connections transmit signals from one tree to the next, helping the trees exchange news about insects, drought, and other dangers. Science has adopted a term first coined by the journal *Nature* for Dr. Simard's discovery of the "wood wide web" pervading our forests.[9] What and how much information is exchanged are subjects we have only just begun to research. For instance, Simard discovered that different tree species are in contact with one another, even when they regard each other as competitors.[10] And the fungi are pursuing their own agendas and appear to be very much in favor of conciliation and equitable distribution of information and resources.[11]

If trees are weakened, it could be that they lose their conversational skills along with their ability to defend themselves. Otherwise, it's difficult to explain why insect pests specifically seek out trees whose health is already compromised. It's conceivable that to do this, insects listen to trees' urgent chemical warnings and then test trees that don't pass the message on by taking a bite out of their leaves or bark. A tree's silence could be because of a serious illness or, perhaps, the loss of its fungal network, which would leave the tree completely cut off from the latest news. The tree no longer registers approaching disaster, and the doors are open for the caterpillar and beetle buffet. The loners I just mentioned are similarly susceptible—they might look healthy, but they have no idea what is going on around them.

In the symbiotic community of the forest, not only trees but also shrubs and grasses—and possibly all plant species—exchange information this way. However, when we step into farm fields, the vegetation becomes very quiet. Thanks to selective breeding, our cultivated plants have, for the most part, lost their ability to communicate above or below ground. Isolated by their silence, they are easy prey for insect pests.[12]

That is one reason why modern agriculture uses so many pesticides. Perhaps farmers can learn from the forests and breed a little more wildness back into their grain and potatoes so that they'll be more talkative in the future.

Communication between trees and insects doesn't have to be all about defense and illness. Thanks to your sense of smell, you've probably picked up on many feel-good messages exchanged between these distinctly different life-forms. I am referring to the pleasantly perfumed invitations sent out by tree blossoms. Blossoms do not release scent at random or to please us. Fruit trees, willows, and chestnuts use their olfactory missives to draw attention to themselves and invite passing bees to sate themselves. Sweet nectar, a sugar-rich liquid, is the reward the insects get in exchange for the incidental dusting they receive while they visit. The form and color of blossoms are signals, as well. They act somewhat like a billboard that stands out against the general green of the tree canopy and points the way to a snack.

So trees communicate by means of olfactory, visual, and electrical signals. (The electrical signals travel via a form of nerve cell at the tips of the roots.) What about sounds? Let's get back to hearing and speech. When I said at the beginning of this chapter that trees are definitely silent, the latest scientific research casts doubt even on this statement. Along with colleagues from Bristol and Florence, Dr. Monica Gagliano from the University of Western Australia has, quite literally, had her ear to the ground.[13] It's not practical to study trees in the laboratory; therefore, researchers substitute grain seedlings because they are easier to handle. They started listening, and it didn't take them long to discover that their measuring apparatus was registering roots crackling quietly at a frequency of 220 hertz. Crackling roots? That doesn't necessarily mean anything. After all, even dead wood crackles when it's burned in a stove. But the noises discovered in the laboratory caused the researchers to sit up and pay attention. For the roots of seedlings not directly involved in the experiment reacted. Whenever the seedlings' roots were exposed to a crackling at 220 hertz, they oriented their tips in that direction. That means the grasses were registering this frequency, so it makes sense to say they "heard" it.

Plants communicating by means of sound waves? That makes me curious to know more, because people also communicate using sound waves. Might this be a key to getting to know trees better? To say nothing of what it would mean if we could hear whether all was well with beeches, oaks, and pines, or whether something was up. Unfortunately, we are not that far advanced, and research in this field is just beginning. But if you hear a light crackling the next time you take a walk in the forest, perhaps it won't be just the wind...

3

— SOCIAL SECURITY —

GARDENERS OFTEN ASK me if their trees are growing too close together. Won't they deprive each other of light and water? This concern comes from the forestry industry. In commercial forests, trees are supposed to grow thick trunks and be harvest-ready as quickly as possible. And to do that, they need a lot of space and large, symmetrical, rounded crowns. In regular five-year cycles, any supposed competition is cut down so that the remaining trees are free to grow. Because these trees will never grow old—they are destined for the sawmill when they are only about a hundred—the negative effects of this management practice are barely noticeable.

What negative effects? Doesn't it sound logical that a tree will grow better if bothersome competitors are removed so that there's plenty of sunlight available for its crown and plenty of water for its roots? And for trees belonging to different species that is indeed the case. They really do struggle with each other for local resources. But it's different for trees of the same species. I've already mentioned that beeches are capable of friendship and go so far as to feed each other. It is obviously not in a forest's best interest to lose its weaker members. If that were to happen, it would leave gaps that would disrupt the forest's sensitive microclimate with its dim light and high humidity. If it weren't for the gap issue, every tree could develop freely and lead its own life. I say "could" because beeches, at least, seem to set a great deal of store by sharing resources.

Students at the Institute for Environmental Research at RWTH Aachen discovered something amazing about photosynthesis in undisturbed beech forests. Apparently, the trees synchronize their performance so that they are all equally successful. And that is not what one would expect. Each beech tree grows in a unique location, and conditions can vary greatly in just a few yards. The soil can be stony or loose. It can retain a great deal of water or almost no water. It can be full of nutrients or extremely barren. Accordingly, each tree experiences different growing conditions; therefore, each tree grows more quickly or more slowly and produces more or less sugar or wood, and thus you would expect every tree to be photosynthesizing at a different rate.

And that's what makes the research results so astounding. The rate of photosynthesis is the same for all the trees. The trees, it seems, are equalizing differences between the strong and the weak. Whether they are thick or thin, all members of the same species are using light to produce the same amount of sugar per leaf. This equalization is taking place underground through the roots. There's obviously a lively exchange going on down there. Whoever has an abundance of sugar hands some over; whoever is running short gets help. Once again, fungi are involved. Their enormous networks act as gigantic redistribution mechanisms. It's a bit like the way social security systems operate to ensure individual members of society don't fall too far behind.[14]

In such a system, it is not possible for the trees to grow too close to each other. Quite the opposite. Huddling together is desirable and the trunks are often spaced no more than 3 feet apart. Because of this, the crowns remain small and cramped, and even many foresters believe this is not good for the trees. Therefore, the trees are spaced out through felling, meaning that supposedly excess trees are removed. However, colleagues from Lübeck in northern Germany have discovered that a beech forest is more productive when the trees are packed together. A clear annual increase in biomass, above all wood, is proof of the health of the forest throng.[15]

When trees grow together, nutrients and water can be optimally divided among them all so that each tree can grow into the best tree it can be. If you "help" individual trees by getting rid of their supposed competition, the remaining trees are bereft. They send messages out to their neighbors in vain, because nothing remains but stumps. Every tree now muddles along on its own, giving rise to great differences in productivity. Some individuals photosynthesize like mad until sugar positively bubbles along their trunk. As a result, they are fit and grow better, but they aren't particularly long-lived. This is because a tree can be only as strong as the forest that surrounds it. And there are now a lot of losers in the forest. Weaker members, who would once have been supported by the stronger ones, suddenly fall behind. Whether the reason for their decline is their location and lack of nutrients, a passing malaise, or genetic makeup, they now fall prey to insects and fungi.

But isn't that how evolution works? you ask. The survival of the fittest? Trees would just shake their heads—or rather their crowns. Their well-being depends on their community, and when the supposedly feeble trees disappear, the others lose as well. When that happens, the forest is no longer a single closed unit. Hot sun and swirling winds can now penetrate to the forest floor and disrupt the moist, cool climate. Even strong trees get sick a lot over the course of their lives. When this happens, they depend on their weaker neighbors for support. If they are no longer there, then all it takes is what would once have been a harmless insect attack to seal the fate even of giants.

In former times, I myself instigated an exceptional case of assistance. In my first years as a forester, I had young trees girdled. In this process, a strip of bark 3 feet wide is removed all around the trunk to kill the tree. Basically, this is a method of thinning, where trees are not cut down, but desiccated trunks remain as standing deadwood in the forest. Even though the trees are still standing, they make more room for living trees, because their leafless crowns allow a great deal of light to reach their neighbors. Do you think this method sounds brutal? I think it does, because death comes slowly over a few years and, therefore, in the future, I wouldn't manage forests this way. I observed how hard the beeches fought and, amazingly enough, how some of them survive to this day.

In the normal course of events, such survival would not be possible, because without bark the tree cannot transport sugar from its leaves to its roots. As the roots starve, they shut down their pumping mechanisms, and because water no longer flows through the trunk up to the crown, the whole tree dries out. However, many of the trees I girdled continued to grow with more or less vigor. I know now that this was only possible with the help of intact neighboring trees. Thanks to the underground network, neighbors took over the disrupted task of provisioning the roots and thus made it possible for their buddies to survive. Some trees even managed to bridge the gap in their bark with new growth, and I'll admit it: I am always a bit ashamed when I see what I wrought back then. Nevertheless, I have learned from this just how powerful a community of trees can be. "A chain is only as strong as its weakest link." Trees could have come up with this old craftsperson's saying. And because they know this intuitively, they do not hesitate to help each other out.

4

— LOVE —

THE LEISURELY PACE at which trees live their lives is also apparent when it comes to procreation. Reproduction is planned at least a year in advance. Whether tree love happens every spring depends on the species. Whereas conifers send their seeds out into the world at least once a year, deciduous trees have a completely different strategy. Before they bloom, they agree among themselves. Should they go for it next spring, or would it be better to wait a year or two? Trees in a forest prefer to bloom at the same time so that the genes of many individual trees can be well mixed. Conifers and deciduous trees agree on this, but deciduous trees have one other factor to consider: browsers such as wild boar and deer.

Boar and deer are extremely partial to beechnuts and acorns, both of which help them put on a protective layer of fat for winter. They seek out these nuts because they contain up to 50 percent oil and starch—more than any other food. Often whole areas of forest are picked clean down to the last morsel in the fall so that, come spring, hardly any beech and oak seedlings sprout. And that's why the trees agree in advance. If they don't bloom every year, then the herbivores cannot count on them. The next generation is kept in check because over the winter the pregnant animals must endure a long stretch with little food, and many of them will not survive. When the beeches or oaks finally all bloom at the same time and set fruit, then it is not possible for the few herbivores left to demolish everything, so there are always enough undiscovered seeds left over to sprout.

"Mast years" is an old term used to describe years when beeches and oaks set seed. In these years of plenty, wild boar can triple their birth rate because they find enough to eat in the forests over the winter. In earlier times, European peasants used the windfall for the wild boar's tame relatives, domestic pigs, which they herded into the woods. The idea was that the herds of domestic pigs would gorge on the wild nuts and fatten up nicely before they were slaughtered. The year following a mast year, wild boar numbers usually crash because the beeches and oaks are taking a time-out and the forest floor is bare once again.

When beeches and oaks put blooming on hold for a number of years, this has grave consequences for insects, as well—especially for bees. It's the same for bees as it is for wild boar: a multi-year hiatus causes their populations to collapse. Or, more accurately, could cause them to collapse, because bees never build up large populations in deciduous forests in the first place. The reason is that true forest trees couldn't care less about these little helpers. What use are the few pollinators left after barren years when you then unfurl millions upon millions of blossoms over hundreds of square miles? If you are a beech or an oak, you have to come up with a more reliable method of pollination, perhaps even one that doesn't exact payment. And what could be more natural than using the wind? Wind blows the powdery pollen out of the blossoms and carries it over to neighboring trees. The wind has a further advantage. It still blows when temperatures fall, even when they drop below 53 degrees Fahrenheit, which is when it gets too chilly for bees and they stay home.

Conifers bloom almost every year, which means bees are an option for pollination because they would always find food. However, conifers are native to northern forests, which are too chilly for bees to be out and about while the trees are blooming, and that is probably why conifers, like beeches and oaks, prefer to rely on the wind. Conifers don't need to worry about taking breaks from blooming, like beeches or oaks, because they have no reason to fear deer and wild boar. The small seeds inside the cones of Spruce & Co. just don't offer an attractive source of nutrition. True, there are birds such as red crossbills, which pick off cones with the tips of their powerful crossed bills and eat the seeds inside, but in general, birds don't seem to be a big problem. And because there is almost no animal that likes to store conifer seeds for winter food, the trees release their potential heirs into the world on tiny wings. Thus equipped, their seeds float slowly down from the tips of their branches and can easily be carried away on a breath of wind.

Spruce & Co. produce huge quantities of pollen, almost as though they wanted to outdo deciduous trees in the mating department. They produce such huge quantities that even in a light breeze, enormous dusty clouds billow over coniferous forests in bloom, giving the impression of a fire smoldering beneath the treetops. This raises the inevitable question about how inbreeding can be avoided in such chaotic conditions. Trees have survived until today only because there is a great deal of genetic diversity within each species. If they all release their pollen at the same time, then the tiny grains of pollen from all the trees mix together and drift through the canopy. And because a tree's own pollen is particularly concentrated around its own branches, there's a real danger its pollen will end up fertilizing its own female flowers. But, as I just mentioned, that is precisely what the trees want to avoid. To reduce this possibility, trees have come up with a number of different strategies.

Some species—like spruce—rely on timing. Male and female blossoms open a few days apart so that, most of the time, the latter will be dusted with the foreign pollen of other spruce. This is not an option for trees like bird cherries, which rely on insects. Bird cherries produce male and female sex organs in the same blossom, and they are one of the few species of true forest trees that allow themselves to be pollinated by bees. As the bees make their way through the whole crown, they cannot help but spread the tree's own pollen. But the bird cherry is alert and senses when the danger of inbreeding looms. When a pollen grain lands on a stigma, its genes are activated and it grows a delicate tube down to the ovary in search of an egg. As it is doing this, the tree tests the genetic makeup of the pollen and, if it matches its own, blocks the tube, which then dries up. Only foreign genes, that is to say, genes that promise future success, are allowed entry to form seeds and fruit. How does the bird cherry distinguish between "mine" and "yours"? We don't know exactly. What we do know is that the genes must be activated, and they must pass the tree's test. You could say, the tree can "feel" them. You might say that we, too, experience the physical act of love as more than just the secretions of neurotransmitters that activate our bodies' secrets, though what mating feels like for trees is something that will remain in the realm of speculation for a long time to come.

Some species have a particularly effective way of avoiding inbreeding: each individual has only one gender. For example, there are both male and female willows, which means they can never mate with themselves but only procreate with other willows. But willows, it must be said, aren't true forest trees. They colonize pioneer sites, areas that are not yet forested. Because there are thousands of wild flowers and shrubs blooming in such places, and they attract bees, willows, like bird cherries, also rely on insects for pollination. But here a problem arises. The bees must first fly to the male willows, collect pollen there, and then transport the pollen to the female trees. If it was the other way around, there would be no fertilization. How does a tree manage this if both sexes have to bloom at the same time? Scientists have discovered that all willows secrete an alluring scent to attract bees. Once the insects arrive in the target area, the willows switch to visual signals. With this in mind, male willows put a lot of effort into their catkins and make them bright yellow. This attracts the bees to them first. Once the bees have had their first meal of sugary nectar, they leave and visit the inconspicuous greenish flowers of the female trees.[16]

Inbreeding as we know it in mammals—that is to say, breeding between populations that are related to one another—is, of course, still possible in all three cases I have mentioned. And here, wind and bees come into play equally. As both bridge large distances, they ensure that at least some of the trees receive pollen from distant relations, and so the local gene pool is constantly refreshed. However, completely isolated stands of rare species of trees, where only a few trees grow, can lose their genetic diversity. When they do, they weaken and, after a few centuries, they disappear altogether.

5

— THE TREE LOTTERY —

TREES MAINTAIN AN inner balance. They budget their strength carefully, and they must be economical with energy so that they can meet all their needs. They expend some energy growing. They must lengthen their branches and widen the diameter of their trunks to support their increasing weight. They also hold some energy in reserve so that they can react immediately and activate defensive compounds in their leaves and bark if insects or fungi attack. Finally, there is the question of propagation.

Species that blossom every year plan for this Herculean task by carefully calibrating their energy levels. However, species that blossom only every three to five years, such as beeches or oaks, are thrown off kilter by such events. Most of their energy has already been earmarked for other tasks, but they need to produce such enormous numbers of beechnuts and acorns that everything else must now take second place. The battle for the branches begins. There's not a speck of space for the blossoms, so a corresponding number of leaves must vacate their posts. In the years when the leaves shrivel and fall off, the trees look unusually bare, so it's no surprise that reports on the condition of forests where the affected trees are growing describe the tree canopy as being in a pitiful state. Because all the trees are going through this process at the same time, to a casual observer the forest looks sick. The forest is not sick, but it is vulnerable. The trees use the last of their energy reserves to produce the mass of blossoms, and to compound the problem, they are left with fewer leaves, so they produce less sugar than they normally do. Furthermore, most of the sugar they do produce is converted into oil and starch in the seeds, so there is hardly any left over for the trees' daily needs and winter stores—to say nothing of the energy reserves intended to defend against sickness.

Many insects have been waiting for just this moment. For example, the beech leaf-mining weevil lays millions upon millions of eggs in the fresh, defenseless foliage. Here, the tiny larvae eat away flat tunnels between the top and bottom surfaces of the leaves, leaving brown papery trails as they feed. The adult beetles chew holes in the leaves until they look as though a hunter has blasted them with a shotgun. Some years, the infestations are so severe that, from afar, the beeches look more brown than green. Normally, the trees would fight back by making the insects' meal extremely bitter—literally. But after producing all those blossoms, they are out of steam, and so this season they must endure the attack without responding.

Healthy trees get over this, especially because afterward there will be a number of years for them to recover. However, if a beech tree is already sickly before the attack, then such an infestation can sound its death knell. Even if the tree knew this, it would not produce fewer blossoms. We know from times of high forest mortality that it is usually the particularly battered individuals that burst into bloom. If they die, their genetic legacy might disappear, and so they probably want to reproduce right away to make sure it continues. Something similar happens after unusually hot summers. After extreme droughts bring many trees to the brink of death, they all bloom together the following year, which goes to show that large quantities of beechnuts and acorns don't indicate that the next winter will be particularly harsh. As blossoms are set the summer before, the abundance of fruit reflects what happened the previous year and has nothing to do with what will happen in the future. The effect of weak defenses shows up again in the fall, this time in the seeds. The beech leaf miners bore into fruit buds as well as leaves. Consequently, although beechnuts form, they remain empty, and therefore, they are barren and worthless.

When a seed falls from a tree, each species has its own strategy as to when the seed sprouts. So how does that work? If a seed lands on soft, damp soil, it has no choice but to sprout as soon as it is warmed by the sun in the spring, for every day the embryonic tree lies around on the ground unprotected it is in great danger—come spring, wild boar and deer are always hungry. And this is just what the large seeds of species such as beeches and oaks do. The next generation emerges from beechnuts and acorns as quickly as it can so that it is less attractive to herbivores. And because this is their one and only plan, the seeds don't have long-term defense strategies against fungi and bacteria. The seeds slough off their protective casings, which lie around on the forest floor through the summer and rot away by the following spring.

Many other species, however, give their seeds the opportunity to wait one or more years until they start to grow. Of course, this means a higher risk of being eaten, but it also offers substantial advantages. For example, seedlings can die of thirst in a dry spring, and when that happens, all the energy put into the next generation is wasted. Or when a deer has its territory and main feeding ground in exactly the spot where the seed lands, it takes no more than a few days for the seedling's tasty new leaves to end up in the deer's stomach. In contrast, if some of the seeds do not germinate for a year or more, then the risk is spread out so that at least a few little trees are likely to make it.

Bird cherries adopt this strategy: their seeds can lie dormant for up to five years, waiting for the right time to sprout. This is a good strategy for this typical pioneer species. Beechnuts and acorns always fall under their mother trees, so the seedlings grow in a predictable, pleasant forest microclimate, but little bird cherries can end up anywhere. Birds that gobble the tart fruit make random deposits of seeds wrapped in their own little packages of fertilizer. If a package lands out in the open in a year when the weather is extreme, temperatures will be hotter and water supplies scarcer than in the cool, damp shadows of a mature forest. Then it's advantageous if at least some of the stowaways wait a few years before waking to their new life.

And after they wake? What are the youngsters' chances of growing up and producing another generation? That's a relatively easy calculation to make. Statistically speaking, each tree raises exactly one adult offspring to take its place. For those that don't make it, seeds may germinate and young seedlings may vegetate for a few years, or even for a few decades, in the shadows, but sooner or later, they run out of steam. They are not alone. Dozens of offspring from other years also stand at their mothers' feet, and by and by, most give up and return to humus. Eventually, a few of the lucky ones that have been carried to open spaces on the forest floor by the wind or by animals get a good start in life and grow to adulthood.

Back to the odds. Every five years, a beech tree produces at least thirty thousand beechnuts (thanks to climate change, it now does this as often as every two or three years, but we'll put that aside for the moment). It is sexually mature at about 80 to 150 years of age, depending on how much light it gets where it's growing. Assuming it grows to be 400 years old, it can fruit at least sixty times and produce a total of about 1.8 million beechnuts. From these, exactly one will develop into a full-grown tree—and in forest terms, that is a high rate of success, similar to winning the lottery. All the other hopeful embryos are either eaten by animals or broken down into humus by fungi or bacteria.

Using the same formula, let's calculate the odds that await tree offspring in the least favorable circumstances. Let's consider the poplar. The mother trees each produce up to 54 million seeds—every year.[17] How their little ones would love to change places with the beech tree youngsters. For until the old ones hand over the reins to the next generation, they produce more than a billion seeds. Wrapped in their fluffy packaging, these seeds strike out via airmail in search of new pastures. But even for them, based purely on statistics, there can be only one winner.

6

— SLOWLY DOES IT —

FOR A LONG time, even I did not know how slowly trees grew. In the forest I manage, there are beeches that are between 3 and 7 feet tall. In the past, I would have estimated them to be ten years old at most. But when I began to investigate mysteries outside the realm of commercial forestry, I took a closer look.

An easy way to estimate the age of a young beech tree is to count the small nodes on its branches. These nodes are tiny swellings that look like a bunch of fine wrinkles. They form every year underneath the buds, and when these grow the following spring and the branch gets longer, the nodes remain behind. Every year, the same thing happens, and so the number of nodes corresponds with the age of the tree. When the branch gets thicker than about a tenth of an inch, the nodes disappear into the expanding bark.

When I examined one of my young beech trees, it turned out that a single 8-inch-long twig already had twenty-five of these swellings. I could find no other indicator of the tree's age on its tiny trunk, which was no more than a third of an inch in diameter, but when I carefully extrapolated the age of the tree from the age of the branch, I discovered that the tree must have been at least eighty years old, maybe more. That seemed unbelievable at the time, until I continued my investigations into ancient forests. Now I know: it is absolutely normal.

Young trees are so keen on growing quickly that it would be no problem at all for them to grow about 18 inches taller per season. Unfortunately for them, their own mothers do not approve of rapid growth. They shade their offspring with their enormous crowns, and the crowns of all the mature trees close up to form a thick canopy over the forest floor. This canopy lets only 3 percent of available sunlight reach the ground and, therefore, their children's leaves. Three percent—that's practically nothing. With that amount of sunlight, a tree can photosynthesize just enough to keep its own body from dying. There's nothing left to fuel a decent drive upward or even a thicker trunk. And rebellion against this strict upbringing is impossible, because there's no energy to sustain it. Upbringing? you ask. Yes, I am indeed talking about a pedagogical method that ensures the well-being of the little ones. And I didn't just come up with the term out of the blue—it's been used by generations of foresters to refer to this kind of behavior.

The method used in this upbringing is light deprivation. But what purpose does this restriction serve? Don't parents want their offspring to become independent as quickly as possible? Trees, at least, would answer this question with a resounding no, and recent science backs them up. Scientists have determined that slow growth when the tree is young is a prerequisite if a tree is to live to a ripe old age. As people, we easily lose sight of what is truly old for a tree, because modern forestry targets a maximum age of 80 to 120 years before plantation trees are cut down and turned into cash.

Under natural conditions, trees that age are no thicker than a pencil and no taller than a person. Thanks to slow growth, their inner woody cells are tiny and contain almost no air. That makes the trees flexible and resistant to breaking in storms. Even more important is their heightened resistance to fungi, which have difficulty spreading through the tough little trunks. Injuries are no big deal for such trees, either, because they can easily compartmentalize the wounds—that is to say, close them up by growing bark over them—before any decay occurs.

A good upbringing is necessary for a long life, but sometimes the patience of the young trees is sorely tested. As I mentioned in chapter 5, "Tree Lottery," acorns and beechnuts fall at the feet of large "mother trees." Dr. Suzanne Simard, who helped discover maternal instincts in trees, describes mother trees as dominant trees widely linked to other trees in the forest through their fungal-root connections. These trees pass their legacy on to the next generation and exert their influence in the upbringing of the youngsters.[18] "My" small beech trees, which have by now been waiting for at least eighty years, are standing under mother trees that are about two hundred years old—the equivalent of forty-year-olds in human terms. The stunted trees can probably expect another two hundred years of twiddling their thumbs before it is finally their turn. The wait time is, however, made bearable. Their mothers are in contact with them through their root systems, and they pass along sugar and other nutrients. You might even say they are nursing their babies.

You can observe for yourself whether young trees are playing the waiting game or putting on a growth spurt. Take a look at the branches of a small silver fir or beech. If the tree is obviously wider than it is tall, then the young tree is in waiting mode. The light it is getting is not sufficient to create the energy it needs to grow a taller trunk, and therefore, the youngster is trying to catch the few leftover rays of sunlight as efficiently as possible. To do this, it lengthens its branches out sideways and grows special ultra-sensitive leaves or needles that are adapted to shade. Often you can't even make out the main shoot on trees like these; they resemble flat-topped bonsai.

One day, it's finally time. The mother tree reaches the end of her life or becomes ill. The showdown might take place during a summer storm. As torrents of rain pour down, the brittle trunk can no longer support the weight of several tons of crown, and it shatters. As the tree hits the ground, it snaps a couple of waiting seedlings. The gap that has opened up in the canopy gives the remaining members of the kindergarten the green light, and they can begin photosynthesizing to their hearts' content. Now their metabolism gets into gear, and the trees grow sturdier leaves and needles that can withstand and metabolize bright light.

This stage lasts between one and three years. Once it is over, it's time to get a move on. All the youngsters want to grow now, and only those that go for it and grow straight as an arrow toward the sky are still in the race. The cards are stacked against those free spirits who think they can meander right or left as the mood takes them and dawdle before they stretch upward. Overtaken by their comrades, they find themselves in the shadows once again. The difference is that it is even darker under the leaves of their cohort that has pulled ahead than it was under their mothers. The teenagers use up the greater part of what weak light remains; the stragglers give up the ghost and become humus once again.

Further dangers are lurking on the way to the top. As soon as the bright sunlight increases the rate of photosynthesis and stimulates growth, the buds of those who have shot up receive more sugar. While they were waiting in the wings, their buds were tough, bitter pills, but now they are sweet, tasty treats—at least as far as the deer are concerned. Because of this, some of the young trees fall victim to these herbivores, ensuring the deers' survival over the coming winter, thanks to the additional calories. But as the crowd of trees is enormous, there are still plenty that keep on growing.

Wherever there is suddenly more light, flowering plants also try their luck, including honeysuckle. Using its tendrils, it makes its way up around the little trunks, always twining in a clockwise direction. By coiling itself around the trunk, it can keep up with the growth of the young tree and its flowers can bask in the sun. However, as the years progress, the coiling vine cuts into the expanding bark and slowly strangles the little tree. Now it is a question of timing: Will the canopy formed by the old trees close soon and plunge the little tree into darkness once again? If it does, the honeysuckle will wither away, leaving only scars. But if there is plenty of light for a while longer, perhaps because the dying mother tree was particularly large and so left a correspondingly large gap, then the young tree in the honeysuckle's embrace can be smothered. Its untimely end, though unfortunate for the tree, brings us some pleasure when we craft its bizarrely twisted wood into walking sticks.

The young trees that overcome all obstacles and continue to grow beautifully tall and slender will, however, have their patience tested yet again before another twenty years have passed. For this is how long it takes for the dead mother's neighbors to grow their branches out into the gap she left when she fell. They take advantage of the opportunity to build out their crowns and gain a little additional space for photosynthesis in their old age. Once the upper story grows over, it is dark once again down below. The young beeches, firs, and pines that have put the first half of their journey behind them must now wait once again until one of these large neighbors throws in the towel. That can take many decades, but even though it takes time, in this particular arena, the die has already been cast. All the trees that have made it as far as the middle story are no longer threatened by competitors. They are now the crown princes and princesses who, at the next opportunity, will finally be allowed to grow up.

7

— FOREST ETIQUETTE —

IN THE FOREST, there are unwritten guidelines for tree etiquette. These guidelines lay down the proper appearance for upright members of ancient forests and acceptable forms of behavior. This is what a mature, well-behaved deciduous tree looks like. It has a ramrod-straight trunk with a regular, orderly arrangement of wood fibers. The roots stretch out evenly in all directions and reach down into the earth under the tree. In its youth, the tree had narrow branches extending sideways from its trunk. They died back a long time ago, and the tree sealed them off with fresh bark and new wood so that what you see now is a long, smooth column. Only when you get to the top do you see a symmetrical crown formed of strong branches angling upward like arms raised to heaven. An ideally formed tree such as this can grow to be very old. Similar rules hold for conifers, except that the topmost branches should be horizontal or bent slightly downward.

And what is the point of all this? Deep down inside, do trees secretly appreciate beauty? Unfortunately, I cannot say, but what I can tell you is that there is a good reason for this ideal appearance: stability. The large crowns of mature trees are exposed to turbulent winds, torrential rains, and heavy loads of snow. The tree must cushion the impact of these forces, which travel down the trunk to the roots. The roots must hold out under the onslaught so that the tree doesn't topple over. To avoid this, the roots cling to the earth and to rocks. The redirected power of a windstorm can tear at the base of the trunk with a force equivalent to a weight of 220 tons.[19] If there is a weak spot anywhere in the tree, it will crack. In the worst-case scenario, the trunk breaks off completely and the whole crown tumbles down. Evenly formed trees absorb the shock of buffeting forces, using their shape to direct and divide these forces evenly throughout their structure.

Trees that don't follow the etiquette manual find themselves in trouble. For example, if a trunk is curved, it has difficulties even when it is just standing there. The enormous weight of the crown is not evenly divided over the diameter of the trunk but weighs more heavily on the wood on one side. To prevent the trunk from giving way, the tree must reinforce the wood in this area. This reinforcement shows up as particularly dark areas in the growth rings, which indicate places where the tree has laid down less air and more wood.

Forked trees are even more precarious. In forked trees, at a certain point, two main shoots form, and they continue to grow alongside each other. Each side of the fork creates its own crown, so in a heavy wind, both sides sway back and forth in different directions, putting a great strain on the trunk where the two parted company. If this transition point is in the shape of a tuning fork or *U*, then usually nothing happens. Woe betide the tree, however, that has a fork in the shape of a *V*, with the two sides joining at a narrow angle. The fork always breaks at its narrowest point, where the two sides diverge. Because the break causes the tree distress, it tries to form thick bulges of wood to prevent further damage. Usually, however, this tactic doesn't work, and bacteria-blackened liquid constantly bleeds from the wound. To make matters worse, the place where one side of the fork broke off gathers water, which penetrates the tear in the bark and causes rot. Sooner or later, a forked tree usually breaks apart, leaving the more stable half standing. This half-tree survives for a few more decades but not much longer. The large gaping wound never heals, and fungi begin to devour the tree slowly from the inside out.

Some trees appear to have chosen the banana as a model for their trunks. The lower part sticks out at an angle, and then the trunk seems to have taken a while to orient itself vertically. Trees like this are completely ignoring the manual, but they don't seem to be alone. Often whole sections of a forest are shaped this way. Are the rules of Nature being set aside here? Not at all. It is Nature herself that forces the trees to adopt such growth patterns.

Take, for example, trees on high mountain slopes just below the tree line. In winter, the snow frequently lies many feet deep, and it is often on the move. And not just in avalanches. Even when it is at rest, snow is sliding at a glacial pace down toward the valleys, even though we can't detect the movement with our eyes. And while the snow is doing that, it's bending trees—the young ones, at least. That's not the end of the world for the smallest among them. They just spring back up again without any ill effects after the snow has melted. However, the trunks of half-grown trees already 10 feet or so tall are damaged. In the most severe cases, the trunk breaks. If it doesn't break, it remains at an angle. From this position, the tree tries to get back to vertical. And because a tree grows only from its tip, the lower part remains crooked. The following winter, the tree is once more pressed out of alignment. Next year's growth points vertically once again. If this game continues for a number of years, gradually you get a tree that is bent into the shape of a saber, or curved sword. It is only with increasing age that the trunk thickens and becomes solid enough that a normal amount of snow can no longer wreak havoc. The lower "saber" keeps its shape, while the upper part of the trunk, left undisturbed, is nice and straight like a normal tree.

Something similar can happen to trees even in the absence of snow, though also on hillsides. In these cases, it is sometimes the ground itself that is sliding extremely slowly down to the valley over the course of many years, often at a rate of no more than an inch or two a year. When this happens, the trees slip slowly along with the ground and tilt over while they continue to grow vertically. You can see extreme cases of this in Alaska and Siberia, where climate change is causing the permafrost to thaw. Trees are losing their footing and being thrown completely off balance in the mushy subsoil. And because every individual tree is tipped in a different direction, the forest looks like a group of drunks staggering around. Accordingly, scientists call these "drunken forests."

At the edge of the forest, the rules for straight trunk growth are not quite so strict. Here, light comes in from the side, from a meadow or a lake—places where trees just don't grow. Smaller trees can get out from under larger ones by growing in the direction of the open area. Deciduous trees, in particular, take advantage of this. If they allow their main shoot to grow almost horizontally, they can increase the size of their crowns by up to 30 feet, thanks to their radically angled trunks. Of course, the trees then risk snapping off, especially after a heavy snowfall, when the laws of physics come into play and the lever principle exacts its tribute. Still, a shorter life-span with enough light for procreation is better than no life at all.

Whereas most deciduous trees leap at chances to grab more light, most conifers stubbornly refuse. They vow to grow straight or not at all. And off they go, always opposing gravity, directly up in a vertical direction so that the trunk is perfectly formed and stable. Lateral branches encountering light at the forest's edge are permitted to put on noticeable girth, but that's it. Only the pine has the cheek to greedily redirect its crown toward the light. No wonder the pine is the conifer with the highest rate of breakage because of snow.

PINE

8

— TREE SCHOOL —

THIRST IS HARDER for trees to endure than hunger, because they can satisfy their hunger whenever they want. Like a baker who always has enough bread, a tree can satisfy a rumbling stomach right away using photosynthesis. But even the best baker cannot bake without water, and the same goes for a tree: without moisture, food production stops.

A mature beech tree can send more than 130 gallons of water a day coursing through its branches and leaves, and this is what it does as long as it can draw enough water up from below.[20] However, the moisture in the soil would soon run out if the tree were to do that every day in summer. In the warmer seasons, it doesn't rain nearly enough to replenish water levels in the desiccated soil. Therefore, the tree stockpiles water in winter.

In winter, there's more than enough rain, and the tree is not consuming water, because almost all plants take a break from growing at that time of year. Together with belowground accumulation of spring showers, the stockpiled water usually lasts until the onset of summer. But in many years, water then gets scarce. After a couple of weeks of high temperatures and no rain, forests usually begin to suffer. The most severely affected trees are those that grow in soils where moisture is usually particularly abundant. These trees don't know the meaning of restraint and are lavish in their water use, and it is usually the largest and most vigorous trees that pay the price for this behavior.

In the forest I manage, the stricken trees are usually spruce, which burst not at every seam but certainly along their trunks. If the ground has dried out and the needles high up in the crown are still demanding water, at some point, the tension in the drying wood simply becomes too much for the tree to bear. It crackles and pops, and a tear about 3 feet long opens in its bark. This tear penetrates deep into the tissue and severely injures the tree. Fungal spores immediately take advantage of the tear to invade the innermost parts of the tree, where they begin their destructive work. In the years to come, the spruce will try to repair the wound, but the tear keeps reopening. From some distance away, you can see a black channel streaked with pitch that bears witness to this painful process.

And with that, we have arrived at the heart of tree school. Unfortunately, this is a place where a certain amount of physical punishment is still the order of the day, for Nature is a strict teacher. If a tree does not pay attention and do what it's told, it will suffer. Splits in its wood, in its bark, in its extremely sensitive cambium (the life-giving layer under the bark): it doesn't get any worse than this for a tree. It has to react, and it does this not only by attempting to seal the wound. From then on, it will also do a better job of rationing water instead of pumping whatever is available out of the ground as soon as spring hits without giving a second thought to waste. The tree takes the lesson to heart, and from then on it will stick with this new, thrifty behavior, even when the ground has plenty of moisture—after all, you never know!

It's no surprise that it is spruce growing in areas with abundant moisture that are affected in this way: they are spoiled. Barely half a mile away, on a dry, stony, south-facing slope, things look very different. At first, I had expected damage to the spruce trees here because of severe summer drought. What I observed was just the opposite. The tough trees that grow on this slope are well versed in the practices of denial and can withstand far worse conditions than their colleagues who are spoiled for water. Even though there is much less water available here year round—because the soil retains less water and the sun burns much hotter—the spruce growing here are thriving. They grow considerably more slowly, clearly make better use of what little water there is, and survive even extreme years fairly well.

A much more obvious lesson in tree school is how trees learn to support themselves. Trees don't like to make things unnecessarily difficult. Why bother to grow a thick, sturdy trunk if you can lean comfortably against your neighbors? As long as they remain standing, not much can go wrong. However, every couple of years, a group of forestry workers or a harvesting machine moves in to harvest 10 percent of the trees in commercial forests in Central Europe. And in natural forests, it is the death from old age of a mighty mother tree that leaves surrounding trees without support. That's how gaps in the canopy open up, and how formerly comfortable beeches or spruce find themselves suddenly wobbling on their own two feet—or rather, on their own root systems. Trees are not known for their speed, and so it takes three to ten years before they stand firm once again after such disruptions.

The process of learning stability is triggered by painful micro-tears that occur when the trees bend way over in the wind, first in one direction and then in the other. Wherever it hurts, that's where the tree must strengthen its support structure. This takes a whole lot of energy, which is then unavailable for growing upward. A small consolation is the additional light that is now available for the tree's own crown, thanks to the loss of its neighbor. But, here again, it takes a number of years for the tree to take full advantage of this. So far, the tree's leaves have been adapted for low light, and so they are very tender and particularly sensitive to light. If the bright sun were to shine directly on them now, they would be scorched—ouch, that hurts! And because the buds for the coming year are formed the previous spring and summer, it takes a deciduous tree at least two growing seasons to adjust. Conifers take even longer, because their needles stay on their branches for up to ten years. The situation remains tense until all the green leaves and needles have been replaced.

The thickness and stability of a trunk, therefore, build up as the tree responds to a series of aches and pains. In a natural forest, this little game can be repeated many times over the lifetime of a tree. Once the gap opened by the loss of another tree is overcome and everyone has extended their crowns so far out that the window of light into the forest is, once again, closed, then everyone can go back to leaning on everyone else. When that happens, more energy is put into growing trunks tall instead of wide, with predictable consequences when, decades later, the next tree breathes its last.

So, let's return to the idea of school. If trees are capable of learning (and you can see they are just by observing them), then the question becomes: Where do they store what they have learned and how do they access this information? After all, they don't have brains to function as databases and manage processes. It's the same for all plants, and that's why some scientists are skeptical and why many of them banish to the realm of fantasy the idea of plants' ability to learn. But, once again, along comes the Australian scientist Dr. Monica Gagliano.

Gagliano studies mimosas, also called "sensitive plants." Mimosas are tropical creeping herbs. They make particularly good research subjects, because it is easy to get them a bit riled up and they are easier to study in the laboratory than trees are. When they are touched, they close their feathery little leaves to protect themselves. Gagliano designed an experiment where individual drops of water fell on the plants' foliage at regular intervals. At first, the anxious leaves closed immediately, but after a while, the little plants learned there was no danger of damage from the water droplets. After that, the leaves remained open despite the drops. Even more surprising for Gagliano was the fact that the mimosas could remember and apply their lesson weeks later, even without any further tests.[21]

It's a shame you can't transport entire beeches or oaks into the laboratory to find out more about learning. But, at least as far as water is concerned, there is research in the field that reveals more than just behavioral changes: when trees are really thirsty, they begin to scream. If you're out in the forest, you won't be able to hear them, because this all takes place at ultrasonic levels. Scientists at the Swiss Federal Institute for Forest, Snow, and Landscape Research recorded the sounds, and this is how they explain them: Vibrations occur in the trunk when the flow of water from the roots to the leaves is interrupted. This is a purely mechanical event and it probably doesn't mean anything.[22] And yet?

We know how the sounds are produced, and if we were to look through a microscope to examine how humans produce sounds, what we would see wouldn't be that different: the passage of air down the windpipe causes our vocal cords to vibrate. When I think about the research results, in particular in conjunction with the crackling roots I mentioned earlier, it seems to me that these vibrations could indeed be much more than just vibrations—they could be cries of thirst. The trees might be screaming out a dire warning to their colleagues that water levels are running low.

9

UNITED WE STAND, DIVIDED WE FALL

TREES ARE VERY social beings, and they help each other out. But that is not sufficient for successful survival in the forest ecosystem. Every species of tree tries to procure more space for itself, to optimize its performance, and, in this way, to crowd out other species. After the fight for light, it is the fight for water that finally decides who wins. Tree roots are very good at tapping into damp ground and growing fine hairs to increase their surface area so that they can suck up as much water as possible. Under normal circumstances, that is sufficient, but more is always better. And that is why, for millions of years, trees have paired up with fungi.

Fungi are amazing. They don't really conform to the one-size-fits-all system we use to classify living organisms as either animals or plants. By definition, plants create their own food out of inanimate material, and therefore, they can survive completely independently. It's no wonder that green vegetation must sprout on barren, empty ground before ani-

mals can move in, for animals can survive only if they eat other living things. Incidentally, neither grass nor young trees like it very much when cattle or deer munch on them. Whether it's a wolf ripping apart a wild boar or a deer eating an oak seedling, in both cases there is pain and death. Fungi are in between animals and plants. Their cell walls are made of chitin—a substance never found in plants—which makes them more like insects. In addition, they cannot photosynthesize and depend on organic connections with other living beings they can feed on.

Over decades, a fungus's underground cottony web, known as mycelium, expands. There is a honey fungus in Switzerland that covers almost 120 acres and is about a thousand years old.[23] Another in Oregon is estimated to be 2,400 years old, extends for 2,000 acres, and weighs 660 tons.[24] That makes fungi the largest known living organisms in the world. The two aforementioned giants are not tree friendly; they kill them as they prowl the forest in search of edible tissue. So let's take a look instead at amicable teamwork between fungi and trees. With the help of mycelium of an appropriate species for each tree—for instance, the oak milkcap and the oak—a tree can greatly increase its functional root surface so that it can suck up considerably more water and nutrients. You find twice the amount of life-giving nitrogen and phosphorus in plants that cooperate with fungal partners than in plants that tap the soil with their roots alone.

To enter into a partnership with one of the many thousands of kinds of fungi, a tree must be very open—literally—because the fungal threads grow into its soft root hairs. There's no research into whether this is painful or not, but as it is something the tree wants, I imagine it gives rise to positive feelings. However the tree feels, from then on, the two partners work together. The fungus not only penetrates and envelops the tree's roots, but also allows its web to roam through the surrounding forest floor. In so doing, it extends the reach of the tree's own roots as the web grows out toward other trees. Here, it connects with other trees' fungal partners and roots. And so a network is created, and now it's easy for the trees to exchange vital nutrients (see chapter 3, "Social Security") and even information—such as an impending insect attack.

This connection makes fungi something like the forest Internet. And such a connection has its price. As we know, these organisms—more like animals in many ways—depend on other species for food. Without a supply of food, they would, quite simply, starve. Therefore, they demand payment in the form of sugar and other carbohydrates, which their partner tree has to deliver. And fungi are not exactly dainty in their requirements. They demand up to a third of the tree's total food production in return for their services.[25] It makes sense, in a situation where you are so dependent on another species, to leave nothing to chance. And so the delicate fibers begin to manipulate the root tips they envelop. First, the fungi listen in on what the tree has to say through its underground structures. Depending on whether that information is useful for them, the fungi begin to produce plant hormones that direct the tree's cell growth to their advantage.[26]

In exchange for the rich sugary reward, the fungi provide a few complimentary benefits for the tree, such as filtering out heavy metals, which are less detrimental to the fungi than to the tree's roots. These diverted pollutants turn up every fall in the pretty fruiting bodies we bring home in the form of porcini, cèpe, or bolete mushrooms. No wonder radioactive cesium, which was found in soil even before the nuclear reactor disaster in Chernobyl in 1986, is mostly found in mushrooms.

Medical services are also part of the package. The delicate fungal fibers ward off all intruders, including attacks by bacteria or destructive fellow fungi. Together with their trees, fungi can live to be many hundreds of years old, as long as they are healthy. But if conditions in their environment change, for instance, as a result of air pollution, then they breathe their last. Their tree partner, however, does not mourn for long. It wastes no time hooking up with the next species that settles in at its feet. Every tree has multiple options for fungi, and it is only when the last of these passes away that it is really in trouble.

Fungi are much more sensitive. Many species seek out trees that suit them, and once they have reserved them for themselves, they are joined to them for better or for worse. Species that like only birches or larches, for instance, are called "host specific." Others, such as chanterelles, get along with many different trees: oaks, birches, and spruce. What is important is whether there is still a bit of room underground. And competition is fierce. In oak forests alone, more than a hundred different species of fungi may be present in different parts of the roots of the same tree. From the oaks' point of view, this is a very practical arrangement. If one fungus drops out because environmental conditions change, the next suitor is already at the door.

Researchers have discovered that fungi also hedge their bets. Dr. Suzanne Simard discovered that their networks are connected not only to a specific tree species but also to trees of different species.[27] Simard injected into a birch tree radioactive carbon that moved through the soil and into the fungal network of a neighboring Douglas fir. Although many species of tree fight each other mercilessly above ground and even try to crowd out each other's root systems, the fungi that populate them seem to be intent on compromise. Whether they actually want to support foreign host trees or only fellow fungi in need of help (which these fungi then pass on to their trees) is as yet unclear.

I suspect fungi are a little more forward "thinking" than their larger partners. Among trees, each species fights other species. Let's assume the beeches native to Central Europe could emerge victorious in most forests there. Would this really be an advantage? What would happen if a new pathogen came along that infected most of the beeches and killed them? In that case, wouldn't it be more advantageous if there were a certain number of other species around—oaks, maples, ashes, or firs—that would continue to grow and provide the shade needed for a new generation of young beeches to sprout and grow up? Diversity provides security for ancient forests. Because fungi are also very dependent on stable conditions, they support other species underground and protect them from complete collapse to ensure that one species of tree doesn't manage to dominate.

If things become dire for the fungi and their trees despite all this support, then the fungi can take radical action, as in the case of the pine and its partner *Laccaria bicolor*, or the bicolored deceiver. When there is a lack of nitrogen, the latter releases a deadly toxin into the soil, which causes minute organisms such as springtails to die and release the nitrogen tied up in their bodies, forcing them to become fertilizer for both the trees and the fungi.[28]

I have introduced you to the most important tree helpers; however, there are many more. Consider the woodpeckers. I wouldn't call them real helpers, but they are of at least some benefit to trees. When bark beetles infest spruce, for example, things get dicey. The tiny insects multiply so rapidly they can kill a tree very quickly by consuming its life-giving cambium layer. If a great spotted woodpecker gets wind of this, it's on the spot right away. Like an oxpecker on a rhinoceros, it climbs up and down the trunk looking for the voracious, fat white larvae. It pecks these out (not thinking particularly of the tree), sending chunks of bark flying. Sometimes this can save the spruce from further damage. Even if the tree doesn't come through this procedure alive, its fellow trees are still protected because now there won't be any adult beetles hatching and flying around. The woodpecker is not in the slightest bit interested in the well-being of the tree, and you can see this particularly clearly in its nesting cavities. It often makes these in healthy trees, severely wounding them as it hacks away. Although the woodpecker frees many trees of pests—for instance, oaks from woodboring beetles—it is more a side effect of its behavior than its intent.

Woodboring beetles can be a threat to thirsty trees in dry years, because the trees are in no position to defend themselves from their attackers. Salvation can come in the form of the black-headed cardinal beetle. In its adult form, it is harmless, feeding on aphid honeydew and plant juices. Its offspring, however, need flesh, and they get this in the form of beetle larvae that live under the bark of deciduous trees. So some oaks have cardinal beetles to thank for their survival. And things can get dire for the beetles as well: once all the children of other species of beetles have been eaten, the larvae turn on their own kind.

10

— THE MYSTERIES OF — MOVING WATER

How does water make its way up from the soil into the tree's leaves? For me, the way this question is answered sums up our current approach to what we know about the forest. For water transport is a relatively simple phenomenon to research—simpler at any rate than investigating whether trees feel pain or how they communicate with one another—and because it appears to be so uninteresting and obvious, university professors have been offering simplistic explanations for decades. This is one reason why I always have fun discussing this topic with students. Here are the accepted answers: capillary action and transpiration.

You can study capillary action every morning at breakfast. Capillary action is what makes the surface of your coffee stand a few fractions of an inch higher than the edge of your cup.

Without this force, the surface of the liquid would be completely flat. The narrower the vessel, the higher the liquid can rise against gravity. And the vessels that transport water in deciduous trees are very narrow indeed: they measure barely 0.02 inches across. Conifers restrict the diameter of their vessels even more, to 0.0008 inches. Narrow vessels, however, are not enough to explain how water reaches the crown of trees that are more than 300 feet tall. In even the narrowest of vessels, there is only enough force to account for a rise of 3 feet at most.[29]

Ah, but we have another candidate: transpiration. In the warmer part of the year, leaves and needles transpire by steadily breathing out water vapor. In the case of a mature beech, the tree exhales hundreds of gallons of water a day. This exhalation causes suction, which pulls a constant supply of water up through the transportation pathways in the tree. Suction works as long as the columns of water are continuous. Bonding forces cause the water molecules to adhere to each other, and because they are strung together like links in a chain, as soon as space becomes available in the leaf thanks to transpiration, the bonded molecules pull each other a little higher up the trunk.

And because even this is not enough, osmosis also comes into play. When the concentration of sugar in one cell is higher than in the neighboring cell, water flows through the cell walls into the more sugary solution until both cells contain the same percentage of water. And when that happens from cell to cell up into the crown, water makes its way up to the top of the tree.

Hmm. When you measure water pressure in trees, you find it is highest shortly before the leaves open up in the spring. At this time of year, water shoots up the trunk with such force that if you place a stethoscope against the tree, you can actually hear it. In the northeastern U.S. and Canada, people make use of this phenomenon to harvest syrup from sugar maples, which are often tapped just as the snow is melting. This is the only time of the year when the coveted sap can be harvested. This early in the year, there are no leaves on deciduous trees, which means there can be no transpiration. And capillary action can be only a partial contributor because the aforementioned rise of 3 feet is hardly worth mentioning. Yet at precisely this time, the trunk is full to bursting. So that leaves us with osmosis, but this seems equally unlikely to me. After all, osmosis works only in the roots and leaves, not in the trunk, which consists not of cells attached one to the other but of long, continuous tubes for transporting water.

So where does that leave us? We don't know. But recent research has discovered something that at least calls into question the effects of transpiration and the forces of cohesion. Scientists from three institutions (the University of Bern; the Swiss Federal Institute for Forest, Snow, and Landscape Research; and the Swiss Federal Institute of Technology in Zurich) listened more closely—literally. They registered a soft murmur in the trees. Above all, at night. At this time of day, most of the water is stored in the trunk, as the crown takes a break from photosynthesis and hardly transpires at all. The trees pump themselves so full of water their trunks sometimes increase in diameter. The water is held almost completely immobile in the inner transportation tubes. Nothing flows. So where are the noises coming from? The researchers think they are coming from tiny bubbles of carbon dioxide in the narrow water-filled tubes.[30] Bubbles in the pipes? That means the supposedly continuous column of water is interrupted thousands of times. And if that is the case, transpiration, cohesion, and capillary action contribute very little to water transport.

So many questions remain unanswered. Perhaps we are poorer for having lost a possible explanation or richer for having gained a mystery. But aren't both possibilities equally intriguing?

11

— TREES AGING GRACEFULLY —

BEFORE I TALK about age, I would like to take a detour into the subject of skin. Trees and skin? First let's approach the subject from the human point of view. Our skin is a barrier that protects our innermost parts from the outer world. It holds in fluids. It stops our insides from falling out. And all the while it releases and absorbs gas and moisture. In addition, it blocks pathogens that would just love to spread through our circulatory system. Aside from that, it is sensitive to contact, which is either pleasant and gives rise to the desire for more, or painful and elicits a defensive response.

Annoyingly, this complicated structure doesn't stay the same forever but gradually sags as we age. Folds and wrinkles appear so that our contemporaries can playfully guess how old we are, give or take a few years. The necessary process of regeneration is not exactly pleasant, either, when looked at close up. Each of us sheds about 0.05 ounces of skin cells a day, which adds up to about a pound a year. The numbers are impressive: 10 billion particles flake off us every day.[31] That doesn't sound very attractive, but sloughing off dead skin is necessary to keep our outer organ in good condition. And in childhood we need to shed skin so that we can grow. Without the ability to renew and expand the covering Nature gives us, sooner or later, we would burst.

And how does this relate to trees? It's just the same with them. The biggest difference is simply the vocabulary we use. The skin of Beeches, Oaks, Spruce & Co. is called bark. It fulfills exactly the same function and protects trees' sensitive inner organs from an aggressive outer world. Without bark, a tree would dry out. And right after the loss of fluid, fungi—which have no chance of survival in healthy, moist wood—would go to town and start breaking everything down. Insects also need lower moisture levels, and if the bark is intact, they are doomed. A tree contains almost as much liquid inside it as we do, and so it's unappealing to pests because they would, quite simply, suffocate.

A break in its bark, then, is at least as uncomfortable for a tree as a wound in our skin is for us. And, therefore, the tree relies on mechanisms similar to the ones we use to stop this from happening. Every year, a tree in its prime adds between 0.5 to 1 inch to its girth. Surely this would make the bark split? It should. To make sure that doesn't happen, the giants constantly renew their skin while shedding enormous quantities of skin cells. In keeping with trees' size in comparison to ours, these flakes are correspondingly larger and measure up to

8 inches across. If you take a look around on the ground under trunks in windy, rainy weather, you will see the remains lying there. The red bark of pines is particularly easy to spot.

But not every tree sheds in the same way. There are species that shed constantly (fastidious people would recommend an anti-dandruff shampoo for such cases). Then there are others that flake with restraint. You can see who's doing what when you look at the exterior of a tree. What you see is the outer layer of bark, which is dead and forms an impervious exterior shell. This outer layer of bark also happens to be a good way of telling different species apart. This works for older trees, anyway, for the distinguishing characteristics have to do with the shapes of the cracks or, you could say, with the folds and wrinkles in a tree's skin. In young trees of all species, the outer bark is as smooth as a baby's bottom. As trees age, wrinkles gradually appear (beginning from below), and they steadily deepen as the years progress. Just how quickly this process plays out depends on the species. Pines, oaks, birches, and Douglas firs start early, whereas beeches and silver firs stay smooth for a long time. It all depends on the speed of shedding.

For beeches, whose silver-gray bark remains smooth until they are two hundred years old, the rate of renewal is very high. Because of this, their skin remains thin and fits their age—that is to say, their girth—exactly and, therefore, doesn't need to crack in order to expand. It's the same for silver firs. Pines and the like, however, drag their feet when it comes to external makeovers. For some reason, they don't like to be parted from their baggage, perhaps because of the additional security a thick skin provides. Whatever the reason, they shed so slowly that they build up really thick outer bark and their exterior layers can be decades old. This means the outer layers originated at a time when the trees were still young and slim, and as the trees age and increase in girth, the outer layers crack way down into the youngest layer of bark that—like the bark of the beeches—fits the girth of the tree as it is now. So, the deeper the cracks, the more reluctant the tree is to shed its bark, and this behavior increases markedly with age.

The same fate catches up with beeches when they pass middle age. This is when their bark starts to get wrinkles, starting from the bottom up. As though they want to broadcast this event, they set to work getting mosses to colonize these nooks and crannies, where moisture from recent rains lingers to soak the plush cushions. You can estimate the age of beech forests from quite a distance: the higher the green growth is up the trunk, the older the trees.

Trees are individuals, and their predisposition to wrinkles varies. Some trees acquire their wrinkles at a younger age than their contemporaries. I have a few beech trees in the forest I manage that at the age of one hundred are covered from top to bottom with rough outer bark. Usually, it takes another 150 years for this to happen. There's no research to show whether this is purely because of genetics or whether a lifetime of excess also plays a role. At least a few factors are, once again, similar to the human condition. The pines in our garden are definitely deeply fissured. This cannot be because of age alone. At about one hundred, they have just outgrown their youth. Since 1934, the year our forester's lodge was built, they have been growing in a particularly sunny spot. Part of the property was cleared to build the lodge, and since then the pines left standing have had more light. More light, more sun, more ultraviolet radiation. The last causes changes in people's skin, and it appears the same thing happens with trees. Intriguingly, the outer bark on the sunny side of the trees is harder, and this means it is more inflexible and more inclined to crack.

The changes I have mentioned, however, can also be because of "skin diseases." In the same way teenage acne often leaves lifelong scars, an attack by bark flies can leave a tree with a rough exterior. In this case, there are no wrinkles; instead, there are thousands of tiny pits and pustules that never disappear no matter how long the tree lives. Sick trees can also develop festering, moist wounds. Bacteria move into these damp areas and stain them black. So, it is not only in people that the skin is a mirror to the soul (or state of well-being).

Old trees can perform another very specific function in the forest ecosystem. In Central Europe, there are no longer any true old-growth forests. The largest extensive stand of trees is between two hundred and three hundred years old. Until these forest preserves become old-growth forests once again, we must look to the West Coast of Canada to understand the role played by ancient trees. There, Dr. Zoë Lindo of McGill University in Montreal researched Sitka spruce that were at least five hundred years old. First of all, she discovered large quantities of moss on the branches and in the branch forks of trees of this advanced age. Blue-green algae had colonized the trees' mossy cushions. These algae capture nitrogen from the air and process it into a form the trees can use. Rain then washes this natural fertilizer down the trunks, making it available to the roots. Thus, old trees fertilize the forest and help their offspring get a better start in life. The youngsters don't have their own moss because moss grows very slowly and takes decades to get established.[32]

Apart from wrinkled skin and mossy growths, there are other physical changes that indicate a tree's age. Take, for example, the crown, which I can compare with something I have as well. Up top, my hair is thinning. It just doesn't grow like it did when I was young. And it's the same with the highest branches up in a tree's crown. After a specific time—one hundred to three hundred years, depending on the species—the annual new growth gets shorter and shorter. In deciduous trees, the successive growth of such short shoots leads to curved, claw-like branches that resemble fingers plagued by arthritis. In conifers, the ramrod-straight trunks end in topmost shoots or leaders that are gradually reduced to nothing. Whereas spruce in this situation stop growing altogether, silver firs continue to grow—but out instead of up, so they look as though a large bird has built its nest in their upper branches. In Germany, where stork nests are a common sight, experts call this phenomenon "stork nest crown." Pines redirect their growth even earlier so that by the time they reach old age, the whole crown is wide with no identifiable leader.

In any event, every tree gradually stops growing taller. Its roots and vascular system cannot pump water and nutrients any higher because this exertion would be too much for the tree. Instead, the tree just gets wider (another parallel to many people of advancing years...). The tree is also not capable of maintaining its mature height for long because its energy levels diminish slowly over the years. At first, it can no longer manage to feed its topmost twigs, and these die off. And so, just as an old person gradually loses body mass, an old tree does too. The next storm sweeps the dead twigs out of the crown, and after this cleanup, the tree looks a little fresher for a while. The process is repeated each year, reducing the crown so gradually we barely notice. Once all the topmost twigs and small branches are lost, only the thicker lower branches remain. Eventually, they die too, though they are not so easily dislodged. Now the tree can no longer hide its advanced age or its infirmity.

It's at this time, if not before, that the bark comes into play once again. Small moist wounds have become portals for fungi to enter. The fungi advertise their triumphant advance through the tree by displaying magnificent fruiting bodies that jut out from the trunk in the shape of semicircular saucers that grow larger with each passing year. Inside, the fungi break down all barriers and penetrate deep into the wood at the heart of the tree. There, depending on the species, they consume stored sugar compounds or, even worse, cellulose and lignin, thereby decomposing the tree's skeleton and reducing it to powder, even though the tree has been bravely resisting this process for decades. On either side of the wounds, which continue to expand, the tree grows new wood, which it builds up into thick stabilizing ridges. For a while, that helps to shore up the ruined structure against powerful winter storms. Then one day, it's all over. The trunk snaps and the tree's life is at an end. "Finally," you can almost hear the young trees-in-waiting sigh. In the years to come, they will quickly push their way up past the crumbling remains. But service in the forest doesn't end when life ends. The rotting cadaver continues to play an important role in the ecosystem for hundreds of years. But more on that later.

12

MIGHTY OAK OR MIGHTY WIMP?

WHEN I WALK through the forest I manage, I often see oaks in distress. And sometimes they are very distressed indeed. Anxious suckers sprouting at the base of the trunk are a dead giveaway. These spindly tufts of growth ring the tree and usually quickly wither away. They indicate that the tree is engaged in an extended fight to the death, and it is panicking. It doesn't make any sense for the oak to attempt to grow leaves so close to the ground, because oaks are trees that need light. They need very bright conditions to photosynthesize. Their ground-hugging solar panels don't produce any energy in the twilight of the understory, and the superfluous arrays are quickly done away with.

A healthy tree doesn't bother to sink energy into developing that kind of growth, preferring instead to extend the reach of its crown up above. At least, that's what it does when it is left in peace. However, oaks in Central European forests are not left in peace, because this is the homeland of the beech. Now, the beech is an amazingly socially oriented tree, but only when it comes to its own kind. Beeches harass other species, such as oaks, to such an extent that they weaken.

It all starts quite slowly and harmlessly when a jay buries a beechnut at the feet of a mighty oak. Because the jay has enough food cached away elsewhere, this beechnut lies undisturbed, and come spring, it sprouts. Slowly, over the course of a number of decades, the sapling grows upward, quietly and unnoticed. True, the young tree doesn't have its mother, but at least the old oak provides shade, and so it helps to raise the youngster at a measured, healthy pace. What looks harmonious above ground turns out to be the beginning of a fight for survival below the surface.

The beech roots penetrate every space the oak is not using, burrowing beneath the old trunk and snapping up water and food the big tree had counted on using for itself. This causes a subtle weakening in the oak. After about 150 years, the little beech tree has grown so tall that it is gradually growing into the crown of the oak. Into, and after a few more decades, through and past, for in contrast to its competitor, the beech can extend its crown and keep growing almost all its life.

By now, the beech leaves are getting direct sunlight, so the tree has all kinds of energy to expand. It grows an impressive crown, which catches 97 percent of the sunlight, just as beech crowns always do. The oak finds itself relegated to the second tier, where its leaves make a vain attempt to snatch some light.

Sugar production is drastically reduced, reserves are used up, and the oak slowly starves.

The oak realizes it cannot beat this stiff competition and will never be able to grow tall shoots to overtake the beech. In its time of need, perhaps in the face of rising panic, it does something that goes against all the rules: it grows new shoots and leaves way down at its base. The leaves are particularly large and soft and can manage with less light than the leaves up in the crown. However, 3 percent is simply not enough. An oak is not a beech. Therefore, these sprouts of anxiety wither, and the precious energy it took to grow them has been squandered. At this stage of starvation, the oak can hold out for a few more decades, but at some point, it gives up. Its powers are waning, and woodboring beetles might come along and put it out of its misery. The beetles lay their eggs under the bark, and the wriggling larvae make short work of feeding on the tree's skin and ending the defenseless tree's life. So, is the mighty oak really a mighty wimp? How did such a weakling of a tree become the symbol for fortitude and longevity?

No matter how badly this tree might fare in most forests in comparison with beeches, an oak can be very tough if it doesn't have any competition. Consider open spaces—specifically our cultivated landscape. Whereas beeches last barely more than two hundred years outside the cozy atmosphere of their native forests, oaks growing near old farmyards or out in pastures easily live for more than five hundred. And what if an oak gets a deep wound or a wide crack in its trunk as a result of a lightning strike? That doesn't matter to the oak, because its wood is permeated with substances that discourage fungi and severely slow down fungal decomposition. These tannins also scare off most insects and, incidentally and inadvertently, improve the taste of wine—should a barrel ever be made from the tree. (Think "oaked" wine.) Even severely damaged trees with major branches broken off can grow replacement crowns and live for a few hundred years longer. Most beech trees wouldn't be able to do that, and they certainly wouldn't be able to do it outside the forest without their beloved networked connections. A storm-battered beech is able to hang on for no more than a couple of decades.

In the forest I manage, oaks show they are made from very stern stuff. On a particularly warm south-facing slope, there are quite a few oaks clawing at naked rock with their roots. When the summer sun heats the stones unbearably, the last drops of water evaporate. In winter, bone-chilling frost penetrates deeply in the absence of the thick protective layer of earth mixed with copious quantities of rotting leaves that you find on the forest floor. The slightest wind catches the slope, so only a few meager lichens grow there, and they do little to moderate temperature extremes. The result? After a century, the trees, or rather miniature trees, are no thicker than your wrist and barely more than 15 feet tall. While their colleagues have grown substantial trunks and more than 100 feet tall in the cozy environment of the forest, these frugal trees make do and content themselves with growing no taller than shrubs. But they do survive! The advantage of this hardscrabble existence is that other species gave up long ago. So, it seems, there can be advantages to a life of deprivation if it means you don't have to worry about competition from other trees.

The thick outer layer of the oak's bark is also much more robust than the smooth, thin skin of the beech, and it can take a great deal of punishment. This has given rise to a saying in German, *"Was schert es eine alte Eiche, wenn sich ein Wildschwein an ihr scheuert?"* which roughly translates as: "It's no skin off an old oak's back if a wild boar wants to use its bark as a scratching post."

13

— SPECIALISTS —

TREES CAN GROW in many extreme environments. Can? Indeed, they must! When a seed falls from a tree, its landing site can be changed only if the wind blows or an animal moves it. And once it has sprouted in the spring, the die is cast. From that point forward, the seedling is bound to this little piece of earth for the rest of its life and must take whatever life hands out. And for most tree youngsters, life hands out a whole series of challenges, for the place where a seed happens to end up often turns out to be highly unsuitable.

It's either too dark, as it is when a light-hungry bird cherry sprouts under large beeches. Or it's too bright, which is the case for beech youngsters whose delicate foliage gets scorched by the blazing sun in clearings. Marshy forest floors rot the roots of most trees, whereas in dry, sandy soil they die of thirst. Places with no nourishing soil at all—rocks or branch forks in big trees—are particularly unfortunate landing sites.

And sometimes luck doesn't last. Consider seeds that come to rest in the tall stumps left when trees snap. The settled seeds grow into little trees whose roots descend into the moldering wood. But when the first unusually dry summer rolls around, causing the last of the moisture to evaporate even from decaying wood, the would-be winners wither and die.

Many Central European tree species have similar ideas about the ideal place to live, because similar criteria for well-being hold true for most of them. They love nutrient-rich, loose, crumbly soil that is well aerated to a depth of many feet. The ground should be nice and moist, especially in summer. But it shouldn't get too hot, and in winter, it shouldn't freeze too much. Snowfall should be moderate but sufficient that when the snow melts, it gives the soil a good soaking. Fall storms should be moderated by sheltering hills or mountain ridges, and the forest shouldn't harbor too many fungi or insects that attack bark or wood.

If trees could dream of an earthly paradise, this is what it would look like. But apart from a few small pockets, these ideal conditions are nowhere to be found. And that is a good thing for species diversity. If Central Europe were such a paradise, the competition would be won almost exclusively by beeches. They know exactly how to exploit abundance, and they suppress competitors by growing up through the crowns of other trees and then covering the losers with their upper branches. If a tree is going to survive such powerful competition, it has to come up with an alternative strategy, but deviations from the arboreal idea of paradise make life difficult for trees, and any tree that wants to find an ecological niche next to a beech must be ready to practice self-denial in one area or another. But are we really talking about ecological niches? As almost no habitat on Earth offers ideal living conditions, it's actually got more to do with the tree adapting than the niche being ideal. There are any number of difficult sites, and a tree that can get along in such places can conquer an enormous geographic range. And that's basically what the spruce has done.

Spruce can gain a foothold everywhere where summers are short and winters are bitterly cold—from the Far North to mountain ranges in Central Europe near the tree line. Because the growing season in Siberia, Canada, and Scandinavia is often only a few weeks long, a beech growing there wouldn't even have a chance to open up all its leaves before the end of the season. And the winter is so bitterly cold that the tree would have frostbite long before it was over. In such regions, it's the spruce that prevails.

Spruce store essential oils in their needles and bark, which act like antifreeze. And that's why they don't need to jettison their green finery but keep it wrapped around their branches in the cold season. As soon as the weather warms up in the spring, they can start photosynthesizing. Not a day is lost, and even if there are only a few weeks in which sugar and wood can be produced, the tree can still grow an inch or two every year.

However, holding on to needles is also extremely risky. Snow lands on the branches and accumulates until the load is so heavy it can break the tree. The spruce employs two defense mechanisms to avoid this. First, a spruce usually grows an absolutely straight trunk. When a structure is nice and vertical, it is difficult to upset its equilibrium. Second, in summer, the branches stick out horizontally. As soon as snow lands on them, they gradually angle down until they are layered one on top of the other like tiles on a roof. Arranged like this, they mutually support each other, and the tree, when viewed from above, presents a much skinnier profile. This means that most of the snow falls around the tree and not on it. Spruce growing in snowy areas at high altitudes or in the Far North also form very long, narrow crowns with short branches, and these slim the trees down even more.

Holding on to needles flirts with yet another danger. Needles increase the surface area the tree presents to the wind, and therefore, spruce are prone to toppling over in winter storms. The only thing that protects them is their extremely slow rate of growth. Trees hundreds of years old are often no taller than 30 feet, and statistically speaking, the danger of being blown over doesn't increase significantly until the trees are more than 80 feet tall.

The natural forest in Central European latitudes is overwhelmingly beech, and beech trees allow very little light to reach the ground. The yew, the epitome of frugality and patience, has decided to make the most of these conditions. Because it knows it can't hold a candle to the beech in the growth department, it has decided to specialize in the forest understory. And here, with the help of the 3 percent of residual light the beeches allow to filter down through their leaves, it grows. Under these conditions, it can take a whole century before a yew reaches 20 to 30 feet and sexual maturity, and a lot can happen to it in this time. Herbivores can nibble it down and set it back by decades. Or worse, a dying beech could knock it over completely. But this tough little tree has taken precautions. Right from the beginning, it puts considerably more energy into building up its root system than other species of trees. Here, it stashes away nutrients, and if misfortune strikes above ground, it grows right back without missing a beat. This often leads to the formation of multiple trunks, which may merge when the tree reaches an advanced age, giving the tree an untidy appearance. And boy can these trees grow old! Living to be a thousand years old or more, they easily outstrip the closest competition, and over the course of centuries, they increasingly get to bask in the sun whenever an old tree growing above them breathes its last. Despite this, yews grow no more than 65 feet tall. They are fine with this, and they don't strive to reach greater heights.

The hornbeam (which, though you wouldn't know it from its name, is related to the birch) tries to imitate the yew, but is not quite so frugal in its habits and needs a bit more light. But it does survive under the beeches, even though it doesn't grow into a large tree here. A hornbeam rarely grows taller than 65 feet anyway, and it reaches that height only when it grows under trees that allow light through, such as oaks. Here, the hornbeam is free to develop, and as it doesn't get in the way of the larger oaks, at least, there is plenty of room for both species. But often along comes a beech that pulls rank on both of them and grows up and over the oaks. The hornbeam can compete only where there is not only a great deal of shade but also severe drought and heat. Here, beeches have to give up eventually, which means, on dry south-facing slopes at least, the hornbeam stands a chance of emerging the winner.

In swampy ground and standing, oxygen-depleted water, the roots of most trees don't survive and the trees die off. You find these conditions near springs or along the banks of streams where the flood plain is regularly underwater. Say a beechnut finds itself there by mistake and sprouts. At first, it might grow into an imposing tree. But sometime during a summer thunderstorm, the tree will fall over when its rotten roots lose their footing. Spruce, pines, hornbeams, and birches run into similar problems when their roots spend some or all of their time in stagnant water. It's completely the opposite for alders. At around 100 feet, it's true they don't grow as tall as their competitors, but they have no problem growing on unpopular swampy ground. Their secret is a system of air ducts inside their roots. These transport oxygen to the tiniest tips, a bit like divers who are connected to the surface via a breathing tube. In addition, the trees have cork cells in the lower parts of their trunks, which allow air to enter. It is only when the water level remains higher than these breathing holes for an extended period of time that the alders weaken sufficiently for their roots to fall victim to aggressive fungi.

14

— TREE OR NOT TREE? —

WHAT IS A tree exactly? The dictionary defines it as a woody plant with a trunk from which branches grow. So the main shoot must be dominant and grow steadily upward or the plant is classified as a shrub, which has many smaller trunks—or rather branches—that originate from a common rootstock. But what about size? Personally, I'm always bothered when I see reports about Mediterranean forests that look to me like a collection of bushes. Trees are, after all, majestic beings, under whose crowns we seem as insignificant as ants in the grass. But then again, on a journey to Lapland, I stumbled upon completely different ambassadors of the tree family that made me feel like Gulliver in Lilliput.

I'm talking about dwarf trees on the tundra, which are sometimes trampled to death by travelers who don't even know they are there. It can take these trees a hundred years to grow just 8 inches tall. I have to say that science doesn't recognize them as trees, and it doesn't accord tree status to the Arctic shrubby birch, either (as you can tell by its name). The latter can grow little trunks up to 10 feet tall, but mostly they remain below eye level and, therefore, are clearly not taken seriously. But if you were to apply the same measure to other trees, then small beeches or mountain ash wouldn't count as trees either. These two are often browsed on so heavily by large mammals such as deer that they grow multiple shoots like bushes and hold out at a height of 20 inches for decades.

And what if you cut a tree down? Is it then dead? What about the centuries-old stump I introduced you to at the beginning of this book that is still alive today, thanks to its comrades? Is that a tree? And, if it isn't, then what is it? It gets even more complicated when a new trunk grows out of an old stump. In many woods, this happens all the time. For centuries in Europe, deciduous trees were cut right down to the base of their trunks by charcoal burners, who harvested them to make charcoal. New trunks grew from the base, forming the foundation for the deciduous woods we have today. Oak and hornbeam forests, in particular, originate from this kind of harvesting, which is known as coppicing. In these forests, the cycle of cutting back and allowing the trees to regrow was repeated every few decades, so the trees never grew tall or matured. Coppicing was popular because people were so poor in those days that they couldn't afford to wait any longer for new wood. You can spot these relics of bygone times when you take a walk in a European forest. Look for trees that have numerous bushy trunks or thick callouses at the base where periodic felling has encouraged a proliferation of growth. Are these trunks now young trees, or alternatively, are they really thousands of years old?

This is a question also asked by scientists, among them a group researching ancient spruce in Dalarna province in Sweden. The oldest spruce in Dalarna has grown a carpet of flat shrubby growth around its single small trunk. All this growth belongs to one tree, and its roots were tested using carbon 14 dating. Carbon 14 is a radioactive carbon that continuously forms in the atmosphere and then gradually decays. This means that the ratio of carbon 14 to other carbon in the atmosphere is always the same. Once carbon 14 is incorporated into inactive biomasses, for instance wood, the process of decay continues unabated, but no new radioactive carbon is accumulated. The lower the amount of radioactive carbon it contains, the older the tissue must be.

Research revealed the spruce to be an absolutely unbelievable 9,550 years old. The individual shoots were younger, but these new growths from the past few centuries were not considered to be stand-alone trees but part of a larger whole.[33] And, I think, quite rightly so. The root is certainly a more decisive factor than what is growing above ground. After all, it is the root that looks after the survival of an organism. It is the root that has withstood severe changes in climatic conditions. And it is the root that has regrown trunks time and time again. It is in the roots that centuries of experience are stored, and it is this experience that has allowed the tree's survival to the present day. As a result of this research on the spruce, a number of scientific schools of thought have been thrown overboard. On the one hand, before this research, no one had any idea that spruce could live for much more than five hundred years; on the other, until then, people had assumed that this conifer first arrived in this part of Sweden two thousand years ago after the ice retreated. For me, this inconspicuous small plant is a symbol for how little we understand about forests and trees and how many wonders we have yet to discover.

So, let's get back to why the roots are the most important part of a tree. Conceivably, this is where the tree equivalent of a brain is located. Brain? you ask. Isn't that a bit farfetched? Possibly, but now we know that trees can learn. This means they must store experiences somewhere, and therefore, there must be some kind of a storage mechanism inside the organism. Just where it is, no one knows, but the roots are the part of the tree best suited to the task. The old spruce in Sweden also shows that what grows underground is the most permanent part of the tree—and where else would it store important information over a long period of time? Moreover, current research shows that a tree's delicate root network is full of surprises.

It is now an accepted fact that the root network is in charge of all chemical activity in the tree. And there's nothing earth shattering about that. Many of our internal processes are also regulated by chemical messengers. Roots absorb substances and bring them into the tree. In the other direction, they deliver the products of photosynthesis to the tree's fungal partners and even route warning signals to neighboring trees. But a brain? For there to be something we would recognize as a brain, neurological processes must be involved, and for these, in addition to chemical messages, you need electrical impulses. And these are precisely what we can measure in the tree, and we've been able to do so since as far back as the nineteenth century. For some years now, a heated controversy has flared up among scientists. Can plants think? Are they intelligent?

In conjunction with his colleagues, František Baluška from the Institute of Cellular and Molecular Botany at the University of Bonn is of the opinion that brain-like structures can be found at root tips. In addition to signaling pathways, there are also numerous systems and molecules similar to those found in animals.[34] When a root feels its way forward in the ground, it is aware of stimuli. The researchers measured electrical signals that led to changes in behavior after they were processed in a "transition zone." If the root encounters toxic substances, impenetrable stones, or saturated soil, it analyzes the situation and transmits the necessary adjustments to the growing tip. The root tip changes direction as a result of this communication and steers the growing root around the critical areas.

Right now, the majority of plant researchers are skeptical about whether such behavior points to a repository for intelligence, the faculty of memory, and emotions. Among other things, they get worked up about carrying over findings in similar situations with animals and, at the end of the day, about how this threatens to blur the boundary between plants and animals. And so what? What would be so awful about that? The distinction between plant and animal is, after all, arbitrary and depends on the way an organism feeds itself: the former photosynthesizes and the latter eats other living beings. Finally, the only other big difference is in the amount of time it takes to process information and translate it into action. Does that mean that beings that live life in the slow lane are automatically worth less than ones on the fast track? Sometimes I suspect we would pay more attention to trees and other vegetation if we could establish beyond a doubt just how similar they are in many ways to animals.

Soil

15

IN THE REALM OF DARKNESS

FOR US HUMANS, soil is more obscure than water, both literally and metaphorically. Whereas it is generally accepted that we know less about the ocean floor than we know about the surface of the moon,[35] we know even less about life in the soil. Sure, there's a wealth of species and facts that have been discovered and that we can read about. But we know only a tiny fraction of what there is to know about the complex life that busies itself under our feet. Up to half the biomass of a forest is hidden in this lower story. Most lifeforms that bustle about here cannot be seen with the naked eye. And that is probably the reason we are not as interested in them as we are in, say, wolves, black woodpeckers, or fire salamanders. For trees, though, these creatures are probably way more important. A forest would have no problem doing without its larger inhabitants. Deer, wild boar, carnivores, and even most birds wouldn't leave any yawning gaps in the ecosystem. Even if they were all to disappear at once, the forest would simply go on growing without many adverse effects. Things are completely different when it comes to the tiny creatures under their feet. There are more life forms in a handful of forest soil than there are people on the planet. A mere teaspoonful contains many miles of fungal filaments. All these work the soil, transform it, and make it so valuable for the trees.

Before we take a closer look at some of these creatures, I'd like to take you back to when soil was first created. Without soil there would be no forests, because trees must have somewhere to put down roots. Naked rock doesn't work, and loosely packed stones, even though they offer roots some support, cannot store sufficient quantities of water or food. Geological processes—such as those active in the ice ages with their sub-zero temperatures—cracked open rocks, and glaciers ground the fragments down into sand and dust until, finally, what was left was a loosely packed substrate. After the ice retreated, water washed this material into depressions and valleys, or storms carried it away and laid it down in layers many tens of feet thick.

Life came along later in the form of bacteria, fungi, and plants, all of which decomposed after death to form humus. Over the course of thousands of years, trees moved into this soil—which only at this stage can be recognized as such—and their presence made it even more precious. Trees stabilized the soil with their roots and protected it against rains and storms. Erosion became a thing of the past, and instead, the layers of humus grew deeper, creating the early stages of bituminous coal. While we are on the subject of erosion: it is one of the forest's most dangerous natural enemies. Soil is lost whenever there are extreme weather events, usually following particularly heavy downpours. If the forest soil cannot absorb all the water right away, the excess runs over the soil surface, taking small particles of soil with it. You can see this for yourself on rainy days: whenever water is brownish in color, this means it is carrying off valuable soil. The forest can lose as much as 2,900 tons per square mile per year. The same area can replace only 290 tons annually through the weathering of stones underground, leading to a huge annual loss of soil. Sooner or later, only the stones remain. Today, you can find many such depleted areas in forests growing in exhausted soils that were cultivated centuries ago. In contrast, forests left undisturbed lose only 1 to 14 tons of soil per square mile per year. In intact forests, the soil under the trees becomes deeper and richer over time so that growing conditions for trees constantly improve.[36]

This brings us to the animals in the soil. Admittedly, they are not particularly attractive. Because of their small size, most species cannot be detected with the naked eye, and even if you go out armed with a magnifying glass, you won't have any luck. It's certainly true that beetle mites, springtails, and pseudocentipedes are not nearly as engaging as orangutans or humpback whales, but in the forest, these little guys are the first link in the food chain and can, therefore, be considered terrestrial plankton. Unfortunately, researchers are only peripherally interested in the thousands of species discovered so far and given unpronounceable Latin names. Countless more species are waiting in vain to be discovered. Perhaps, however, we can take comfort from this: there are still many secrets in the forest that lies directly outside your back door. Let's take a look at the little that has been brought to light so far.

Let's take the aforementioned beetle, or oribatid, mites, of which there are about a thousand known species in European latitudes. They are less than 0.04 inches long and look like spiders with inadvisably short legs. Their bodies are two-tone brown, which blends in well with their natural environment: the soil. Mites? That brings up associations with household dust mites, which feed on the flakes of skin we shed and other waste products and may trigger allergies in some people. At least some of the beetle mites act in a similar way around trees. The leaves and fragments of bark trees shed would pile up several yards deep if it weren't for a hungry army of microscopic creatures ready to pounce on the detritus. To do this, they live in the cast-off leaf litter, which they devour voraciously. Other species specialize in fungi. These creatures crouch in small underground tunnels and suck the juices that ooze out of the fungi's fine white threads. Finally, beetle mites feed on the sugar trees share with their fungal partners. Whether it's rotting wood or dead snails, there is nothing that doesn't have its corresponding beetle mite. They appear everywhere at the intersection between birth and decay, and so they must be considered essential components of the ecosystem.

Then there are the weevils. They look a bit like tiny elephants that have lost their enormous ears, and they belong to the most species-rich family of insects in the world. In Europe alone there are about 1,400 species. For the weevils it's not so much about eating as it is about child care. With the help of their long snouts, the little creatures eat small holes in leaves and stems, where they lay their eggs. Protected from predators, the larvae gnaw little passages inside the plants and grow in peace.[37]

Some species of weevil, mostly those that live on the forest floor, can no longer fly because they have become accustomed to the slow rhythms of the forest and its practically eternal existence. The farthest they can travel is 30 feet a year, and they really don't need to be able to travel any farther than that. If the environment around a tree changes because the tree dies, all a weevil has to do is make it to the next tree and continue nibbling around there in the rotting leaf litter. If you find weevils, you can be sure the forest has a long uninterrupted history. If the forest was cleared in the Middle Ages and later replanted, you won't find these insects, because it would simply have been too far for them to walk to the next old forest.

All the animals I have mentioned so far have one thing in common: they are very small and, therefore, their circle of influence is extremely limited. In the large old-growth forests that once covered Central Europe, this didn't matter at all. Today, however, people have altered most of the forests. There are spruce instead of beeches, Douglas firs instead of oaks, young trees instead of old ones. The new forests were literally no longer to the animals' taste, and so they starved and local populations died out. However, there are still a few old deciduous forests that act as refuges where the original diversity of species still exists. All over Germany, forestry commissions are trying to grow more deciduous than coniferous forests once again. But if mighty beeches are to herald change and stand once again where spruce now topple in storms, how will the beetle mites and springtails get back to these places? Not by walking there, that's for sure, because they cover barely 3 feet in a lifetime. So is there any hope at all that one day, at least in national parks such as the Bavarian Forest, we will once again be able to marvel at authentic old-growth forests? It is entirely possible.

Research carried out by students in the forest I manage has shown that microscopic organisms—at least those associated with coniferous forests—can cover astonishing distances. Old spruce plantations show this particularly clearly. Here, the young researchers found species of springtails that specialize in spruce forests. But my predecessors here in Hümmel planted such forests only a hundred years ago. Prior to that, we had predominantly old beech trees, just like everywhere else in Central Europe. So how did these conifer-dependent springtails get to Hümmel? My guess is that it must have been birds that brought these terrestrial creatures as stowaways in their plumage. Birds love to take dust baths in dead leaves to clean their feathers. When they do this, tiny creatures that live in the soil must surely get trapped, and they are then unloaded during a dust bath in the next forest. And what works for animals specialized for spruce probably also works for species that love deciduous trees. If in the future, more mature deciduous forests are allowed, once again, to develop undisturbed, then birds can see to it that the appropriate sub-letters show up again as well.

In any event, the return of the teeny little creatures can take a very, very long time, as the latest studies out of Kiel and Lüneburg attest.[38] More than a hundred years ago, oak forests were planted on the Lüneburg Heath on what had once been arable land. It would take only a few decades for the original framework of fungi and bacteria to settle the soil once again—or so the scientists assumed. But far from it. Even after this relatively long time, there are still gaping holes in the species' inventory, and this deficit has grave consequences for the forest, as the nutrient cycles of birth and decay aren't functioning properly. Moreover, the soil still contains excess nitrogen from the fertilizers once used there. True, the oak forest is growing more quickly than similar stands of trees located on ancient forest soil, but it is markedly less robust when it comes to issues such as drought. We don't know how long it will take until true forest soil is created once again, but we do know that a hundred years is not enough.

To make it possible for this regeneration to happen at all, you need preserves with ancient forests free from any human interference. These are places where the diversity of soil life can survive, and these refuges can be the nucleuses for recovery in surrounding areas. And, incidentally, no real sacrifices need to be made to make this happen, as the community of Hümmel has demonstrated for years. They have put entire old beech forests under protection and found innovative ways to market them. Part of the forest is used as an arboreal mortuary, where the trees are leased out as living gravestones for urns buried under them. To become part of the ancient forest after death—isn't that a wonderful idea? Another part of the preserve is leased to firms as their contribution to protecting the environment. This makes up for the fact that the wood itself is not being used, and both people and Nature are happy.

Efforts to offset the costs of protecting and restoring forests in the twenty-first century are happening around the world. Some combine utility with education: tourists in the Maya Biosphere Reserve in Guatemala employ residents who would otherwise be cutting down forests to sell the lumber and grow food in the clearings. Some combine prestige with preservation: in Scotland, you can buy a piece of forest originally owned by the nobility to keep lumber companies out and help usher in the return of the ancient Caledonian Forest. Yet others involve unlikely partners: the U.S. Department of Defense contributes to the National Fish and Wildlife Foundation's efforts to restore longleaf pine ecosystems in the American southeast on the grounds that forested buffers around military bases contribute to military readiness.[39] There are so many ways that forests can be kept both undisturbed and productive!

16

— CARBON DIOXIDE VACUUMS —

IN A VERY simple, widely circulated image of natural cycles, trees are poster children for a balanced system. As they photosynthesize, they produce hydrocarbons, which fuel their growth, and over the course of their lives, they store up to 22 tons of carbon dioxide in their trunks, branches, and root systems. When they die, the same exact quantity of greenhouse gases is released as fungi and bacteria break down the wood, process the carbon dioxide, and breathe it out again. The assertion that burning wood is climate neutral is based on this concept. After all, it makes no difference if it's small organisms reducing pieces of wood to their gaseous components or if the home hearth takes on this task, right? But how a forest works is way more complicated than that. The forest is really a gigantic carbon dioxide vacuum that constantly filters out and stores this component of the air.

It's true that some of this carbon dioxide does indeed return to the atmosphere after a tree's death, but most of it remains locked in the ecosystem forever. The crumbling trunk is gradually gnawed and munched into smaller and smaller pieces and worked, by fractions of inches, more deeply into the soil. The rain takes care of whatever is left, as it flushes organic remnants down into the soil. The farther underground, the cooler it is. And as the temperature falls, life slows down, until it comes almost to a standstill. And so it is that carbon dioxide finds its final resting place in the form of humus, which continues to become more concentrated as it ages. In the far distant future, it might even become bituminous or anthracite coal.

Today's deposits of these fossil fuels come from trees that died about 300 million years ago. They looked a bit different—more like 100-foot-tall ferns or horsetail—but with trunk diameters of about 6 feet, they rivaled today's species in size. Most trees grew in swamps, and when they died of old age, their trunks splashed down into stagnant water, where they hardly rotted at all. Over the course of thousands of years, they turned into thick layers of peat that were then overlain with rocky debris, and pressure gradually turned the peat to coal. Thus, large conventional power plants today are burning fossil forests. Wouldn't it be beautiful and meaningful if we allowed our trees to follow in the footsteps of their ancestors by giving them the opportunity to recapture at least some of the carbon dioxide released by power plants and store it in the ground once again?

Today, hardly any coal is being formed because forests are constantly being cleared, thanks to modern forest management practices (aka logging). As a result, warming rays of sunlight reach the ground and help the species living there kick into high gear. This means they consume humus layers even deep down into the soil, releasing the carbon they contain into the atmosphere as gas. The total quantity of climate-changing gases that escapes is roughly equivalent to the amount of timber that has been felled. For every log you burn in your fire at home, a similar amount of carbon dioxide is being released from the forest floor outside. And so carbon stores in the ground below trees in our latitudes are being depleted as fast as they are being formed.[40]

Despite this, you can observe at least the initial stages of coal formation every time you walk in the forest. Dig down into the soil a little until you come across a lighter layer. Up to this point, the upper, darker soil is highly enriched with carbon. If the forest were left in peace from now on, this layer would be the precursor of coal, gas, or oil. At least in larger protected areas, such as the hearts of national parks, these processes continue today uninterrupted. And I'd just like to add that meager layers of humus are not the result only of modern forestry practices: way back when in Europe, Romans and Celts were also industriously cutting back forests and disrupting natural processes.

What sense does it make for trees to constantly remove their favorite food from the system? And all plants do this, not just trees. Even algae out in the oceans extract carbon dioxide from the atmosphere. The carbon dioxide sinks into the muck when plants die, where it is stored in the form of carbon compounds. Thanks to these remains—and the remains of animals, such as the calcium carbonate excreted by coral, which is one of the largest repositories of carbon dioxide on earth—after hundreds of millions of years, an enormously large amount of carbon has been removed from the atmosphere. When the largest coal deposits were formed, in the Carboniferous period, carbon dioxide concentrations were much higher—nine times today's levels—before prehistoric forests, among other factors, reduced carbon dioxide to a level that was still triple the concentration we have today.[41]

Where is the end of the road for our forests? Will they go on storing carbon until someday there isn't any left in the air? This, by the way, is no longer a question in search of an answer, thanks to our consumer society, for we have already reversed the trend as we happily empty out the earth's carbon reservoirs. We are burning oil, gas, and coal as heating materials and fuel, and spewing their carbon reserves out into the air. In terms of climate change, could it perhaps be a blessing that we are liberating greenhouse gases from their underground prisons and setting them free once again? Ah, not so fast. True, there has been a measurable fertilizing effect as the levels of carbon dioxide in the atmosphere have risen. The latest forest inventories document that trees are growing more quickly than they used to. The spreadsheets that estimate lumber production need to be adjusted now that one third more biomass is accruing than a few decades ago. But what was that again? If you are a tree, slow growth is the key to growing old. Growth fueled by hefty additions of excess nitrogen from agricultural operations is unhealthy. And so the tried and tested rule holds true: less (carbon dioxide) is more (life-span).

When I was a student of forestry, I learned that young trees are more vigorous and grow more quickly than old ones.

The doctrine holds to this day, with the result that forests are constantly being rejuvenated. Rejuvenated? That simply means that all the old trees are felled and replaced with newly planted little trees. Only then, according to the current pronouncements of associations of forest owners and representatives of commercial forestry, are forests stable enough to produce adequate amounts of timber to capture carbon dioxide out of the atmosphere and store it. Depending on what tree you are talking about, energy for growth begins to wane from 60 to 120 years of age, and that means it is time to roll out the harvesting machines. Has the ideal of eternal youth, which leads to heated discussions in human society, simply been transferred to the forest? It certainly looks that way, for at 120 years of age, a tree, considered from a human perspective, has barely outgrown its school days.

In fact, past scientific assumptions in this area appear to have gotten ahold of the completely wrong end of the stick, as suggested by a study undertaken by an international team of scientists. The researchers looked at about 700,000 trees on every continent around the world. The surprising result:

the older the tree, the more quickly it grows. Trees with trunks 3 feet in diameter generated three times as much biomass as trees that were only half as wide.[42] So, in the case of trees, being old doesn't mean being weak, bowed, and fragile. Quite the opposite, it means being full of energy and highly productive. This means elders are markedly more productive than young whippersnappers, and when it comes to climate change, they are important allies for human beings. Since the publication of this study, the exhortation to rejuvenate forests to revitalize them should at the very least be flagged as misleading. The most that can be said is that as far as marketable lumber is concerned, trees become less valuable after a certain age. In older trees, fungi can lead to rot inside the trunk, but this doesn't slow future growth one little bit. If we want to use forests as a weapon in the fight against climate change, then we must allow them to grow old, which is exactly what large conservation groups are asking us to do.

17

— WOODY CLIMATE CONTROL —

TREES DO NOT enjoy extreme changes in temperature or moisture, and regional climates do not spare anything, not even large plants. But have you considered whether trees might be able to exert their influence once in a while? My Eureka moment on this subject came in a little forest growing on dry, sandy, nutrient-deficient soil near Bamberg, Germany. Forest specialists once claimed that only pines could flourish here. To avoid creating a bleak monoculture, beeches were also planted so that their leaves could neutralize the acid in the pine needles to make them more palatable to the creatures in the soil. There was no thought of using the deciduous trees for lumber; they were considered to be so-called service trees. But the beeches had no intention of playing a subservient role. After a few decades, they showed what they were made of.

With their annual leaf fall, the beeches created an alkaline humus that could store a lot of water. In addition, the air in this little forest gradually became moister, because the leaves of the growing beeches calmed the air by reducing the speed of the wind blowing through the trunks of the pines. Calmer air meant less water evaporated. More water allowed the beeches to prosper, and one day they grew up and over the tops of the pines. In the meantime, the forest floor and the microclimate had both changed so much that the conditions became more suited to deciduous trees than to the more frugal conifers. This transformation is a good example of what trees can do to change their environment. As foresters like to say, the forest creates its own ideal habitat.

As I have just explained, as far as calming the wind is concerned, this is certainly possible, but what about budgeting water? Well, if hot summer air cannot blow-dry the forest floor because the soil remains deeply shaded and well protected, then that too is possible. In the forest I manage, students from RWTH Aachen discovered just how great the temperature differences can be between a coniferous plantation that is regularly thinned and a beech forest that has been allowed to age naturally. On an extremely hot August day that chased the thermometer up to 98 degrees Fahrenheit, the floor of the deciduous forest was up to 50 degrees cooler than that of the coniferous forest, which was only a couple of miles away. This cooling effect, which meant less water lost, was very clearly because of the biomass, which also contributed shade. The more living and dead wood there is in the forest, the thicker the layer of humus on the ground and the more water is stored in the total forest mass. Evaporation leads to cooling, which, in turn, leads to less evaporation. To put it another way, in summer an intact forest sweats for the same reason people do and with the same result.

Incidentally, you can indirectly observe trees sweating by looking at houses. You often find Christmas trees with intact root balls that people did not want to discard, all nicely planted by the house and in the best of health. They grow and grow, and sooner or later they get much larger than the homeowners anticipated. Usually, they are planted too close to the side of the house, and some of their branches might even extend out over the roof. And this is when you can see something like sweat stains. These are unpleasant enough when we get them under our arms, but for houses there are more than merely visual consequences. The trees sweat so profusely that algae and moss colonize their moist facades and roof tiles. Rainwater, slowed down by the plant growth, no longer drains away so easily, and dislodged patches of moss clog the gutters. The stucco crumbles over the years because of the damp and has to be replaced prematurely. People who park their cars under trees, however, benefit from the way trees even out extremes. When there are freezing temperatures, people who park their cars out in the open have to scrape ice from their windows, whereas cars parked under trees often remain ice free. Apart from the fact that trees can negatively affect the exterior of buildings, I find it fascinating how much spruce and other species influence microclimates in their vicinity. Consider how much greater the influence of an undisturbed forest must be.

Whoever sweats a lot must also drink a lot. And during a downpour, you have the opportunity to observe a tree taking a long swig. Because downpours usually happen at the same time as storms, I can't recommend taking a walk out into the forest. However, if you, like me (often because of work), are outside anyway, then you can observe a fascinating spectacle. Mostly, it is beeches that indulge in such all-out drinking binges. Like many deciduous trees, they angle their branches up. Or you could say, down. For the crown opens its leaves not only to catch sunlight but also to catch water. Rain falls on hundreds of thousands of leaves, and the moisture drips from them down onto the twigs. From there, it runs along the branches, where the tiny streams of water unite into a river that rushes down the trunk. By the time it reaches the lower part of the trunk, the water is shooting down so fast that when it hits the ground, it foams up. During a severe storm, a mature tree can down an additional couple of hundred gallons of water that, thanks to its construction, it funnels to its roots. There, the water is stored in the surrounding soil, where it can help the tree over the next few dry spells.

Spruce and firs can't do this. The crafty firs like to mix in with the beeches, whereas the spruce usually stick together, which means they're often thirsty. Their crowns act like umbrellas, which is really convenient for hikers. If you're caught in a shower and stick close to the trunks, you'll hardly get wet at all, and neither will the trees' roots. Rainfall of up to 2.5 gallons of water per square yard of forest (that's a pretty good downpour) gets completely hung up in the needles and branches. When the clouds clear, this water evaporates and all this precious moisture is lost. Why do spruce do this? They have, quite simply, never learned to adapt to water shortages.

Spruce are comfortable in cold regions where, thanks to the low temperatures, the groundwater hardly ever evaporates. For instance, they like it up in the Alps just below the tree line where particularly heavy downpours ensure that drought is never an issue. There are heavy snowfalls, though, which is why the branches grow out horizontally or angled slightly down so that they can lean on each other for support when the snow piles up. But this means that water doesn't run down the tree, and when spruce are growing in drier areas at lower elevations, then this winter adaptation is of no use to them. The majority of the coniferous forests we have in Central Europe today were planted, and people put the forests in places that made sense to them. In these places, the conifers are always suffering from thirst, and all the while their built-in umbrellas are intercepting one third of the rain that falls and returning it to the atmosphere. Deciduous forests intercept only 15 percent of the rain that falls, which means they are profiting from 15 percent more water than their needle-bedecked colleagues.

BEECH

18

— THE FOREST AS — WATER PUMP

HOW DOES WATER get to the forest, anyway, or—to take one step further back—how does water reach land at all? It seems like such a simple question, but the answer turns out to be rather complicated. For one of the essential characteristics of land is that it is higher than water. Gravity causes water to flow down to the lowest point, which should cause the continents to dry out. The only reason this doesn't happen is thanks to supplies of water constantly dropped off by clouds, which form over the oceans and are blown over land by the wind. However, this mechanism only functions within a few hundred miles of the coast. The farther inland you go, the drier it is, because the clouds get rained out and disappear. When you get about 400 miles from the coast, it is so dry that the first deserts appear. If we depended on just this mechanism for water, life would be possible only in a narrow band around the edge of continents; the interior of land masses would be arid and bleak. So, thank goodness for trees.

Of all the plants, trees have the largest surface area covered in leaves. For every square yard of forest, 27 square yards of leaves and needles blanket the crowns.[43] Part of every rainfall is intercepted in the canopy and immediately evaporates again. In addition, each summer, trees use up to 8,500 cubic yards of water per square mile, which they release into the air through transpiration. This water vapor creates new clouds that travel farther inland to release their rain. As the cycle continues, water reaches even the most remote areas. This water pump works so well that the downpours in some large areas of the world, such as the Amazon Basin, are almost as heavy thousands of miles inland as they are on the coast.

There are a few requirements for the pump to work: from the ocean to the farthest corner, there must be forest. And, most importantly, the coastal forests are the foundations for this system. If they do not exist, the system falls apart. Scientists credit Anastassia Makarieva from Saint Petersburg in Russia for the discovery of these unbelievably important connections.[44] They studied different forests around the world and everywhere the results were the same. It didn't matter if they were studying a rain forest or the Siberian taiga, it was always the trees that were transferring life-giving moisture into land-locked interiors. Researchers also discovered that the whole process breaks down if coastal forests are cleared. It's a bit like if you were using an electrical pump to distribute water and you pulled the intake pipe out of the pond. The fallout is already apparent in Brazil, where the Amazonian rain forest is steadily drying out. Central Europe is within the 400-mile zone and, therefore, close enough to the intake area. Thankfully, there are still forests here, even if they are greatly diminished.

Coniferous forests in the Northern Hemisphere influence climate and manage water in other ways, too. Conifers give off terpenes, substances originally intended as a defense against illness and pests. When these molecules get into the air, moisture condenses on them, creating clouds that are twice as thick as the clouds over non-forested areas. The possibility of rain increases, and in addition, about 5 percent of the sunlight is reflected away from the ground. Temperatures in the area fall. Cool and moist—just how conifers like it. Given this reciprocal relationship between trees and weather, forest ecosystems probably play an important role in slowing down climate change.[45]

For ecosystems in Central Europe, regular rainfall is extremely important because water and forests share an almost unbreakable bond. Streams, ponds—even the forest itself—all these ecosystems depend on providing their inhabitants with as much stability as they can. A good example of an organism that doesn't like a lot of change is the freshwater snail. Depending on the species, it is often less than 0.08 inches long, and it loves cold water. They like it to be no more than 46 degrees Fahrenheit, and for some freshwater snails the reason for this lies in their past: their ancestors lived in the meltwater that drained off the glaciers covering a large part of Europe in the last ice age.

Clean springs in the forest offer such conditions. The water comes out at a constantly cool temperature, for these springs are where groundwater bubbles to the surface. Groundwater is found deep underground, where it is insulated from outside air temperatures, and therefore, it is as cold in summer as it is in winter. Given that we no longer have any glaciers, this is the ideal replacement habitat for today's freshwater snails. But that means the water has to bubble up year round, and this is where the forest comes into play. The forest floor acts as a huge sponge that diligently collects all the rainfall. The trees make sure that the raindrops don't land heavily on the ground but drip gently from their branches. The loosely packed soil absorbs all the water, so instead of the raindrops joining together to form small streams that rush away in the blink of an eye, they remain trapped in the soil. Once the soil is saturated and the reservoir for the trees is full, excess moisture is released slowly and over the course of many years, deeper and deeper into the layers below the surface. It can take decades before the moisture once again sees the light of day. Fluctuations between periods of drought and heavy rain become a thing of the past, and what remains is a constantly bubbling spring.

Although it has to be said, it doesn't always bubble. Often it looks more like a swampy-squishy area, a dark patch on the forest floor seeping toward the nearest little stream. If you take a closer look (and to do that you must get down on your knees), you can make out the tiny rivulets that betray the existence of a spring. Now, to find out whether this is indeed groundwater or just surface water left over from a heavy shower, reach for your thermometer. Less than 48 degrees

Fahrenheit? Then it is indeed a spring. But what kind of a person carries a thermometer around all the time? Another option is to take a walk when there's a hard frost. Puddles and rainwater will be frozen, while water will still be seeping out of a spring. This then is where the freshwater snails call home, and here they enjoy their preferred temperature year round. And it is not only the forest floor that makes this possible. In summer, a microhabitat like this could warm up quickly and overheat the snails. But the leafy canopy throws shadows that block out most of the sun.

The forest offers a similar and even more important service to streams. The water in a stream is susceptible to greater temperature variations than spring water, which is continuously replaced with cool groundwater. These streams contain animals such as salamander larvae and tadpoles, which are just waiting for their life outside the stream to begin. Like the freshwater snails, they need the water to remain cool so that oxygen doesn't escape, but if the water freezes solid, the baby salamanders will die. It's a good thing deciduous trees just happen to solve this problem. In winter, when there is very little warmth in the sun, bare branches allow a lot of warmth to penetrate. The movement of water over the uneven bottom also protects the stream from suddenly freezing. When the sun climbs higher in the sky in late spring and the air is noticeably warmer, the trees unfurl their leaves, closing the blinds and shading the running water. Then in the fall, when temperatures drop once again, the sky reopens above the stream when the trees drop all their leaves. It's tougher for streams that flow under coniferous trees. It's bitterly cold here in winter, and sometimes the water freezes solid. Because it warms up only gradually in spring, this habitat is just not an option for many organisms. But pitch-dark streams like this are rarely found in Nature, because spruce don't like having wet feet and, therefore, usually keep their distance. It's most often plantations that cause this conflict between coniferous forests and the denizens of streams.

The importance of trees for streams continues even after death. When a dead beech falls across a streambed, it lies there for decades. It acts like a small dam and creates tiny pockets of calm water where species that can't tolerate strong currents can hang out. The nondescript larvae of the fire salamander are just such creatures. They look like small newts, except they have feathery gills. They are finely stippled with dark markings and have a yellow spot where their legs meet their bodies. In the cold water of the forest, they lie in wait for the tiny crawfish they love to eat. These little guys need crystal-clear water, and the dead trees look after this as well. Mud and floating debris drop to the bottom of the tiny dammed pools, and because stream flows are so low, it gives bacteria more time to break down harmful substances. And don't worry about that foam that sometimes forms in these pools after heavy rains. What looks like an environmental disaster is, in fact, the result of humic acids that tiny waterfalls have mixed with air until they turn into froth. These acids come from the decomposition of leaves and dead wood and are extremely beneficial for the ecosystem.

When it comes to the creation of small pools, forests in Central Europe have become less dependent on dead trees falling down. They are increasingly getting help from an animal that has recently made a comeback after being nearly eradicated. This animal is the beaver. I have my doubts whether the trees are really happy about this, for this rodent, which can weigh more than 60 pounds, is the lumberjack of the animal world. It takes a beaver one night to bring down a 3-to-4-inch-thick tree. Larger trees are felled over the course of multiple work shifts. What the beaver is after are twigs and small branches, which it uses for food. It stockpiles enormous quantities in its lodge to last the winter, and as the years pass, the lodge grows by many yards. The branches also camouflage the entrances to the tunnels that lead into the lodge. As an added security feature, the beaver builds these entrances underwater so that predators can't get in. The rest of the living space is above water and therefore dry.

Because water levels can fluctuate wildly with the seasons, many beavers also build dams, blocking streams and turning them into large ponds. Beaver ponds slow the flow of water from the forest, and extensive wetlands form in the areas around the dams. Alders and willows like to grow here; beeches, which cannot stand having wet feet, die off. But the upstart trees in the feeding zone around the lodge don't get to grow old, for they are the beavers' living larder. Although beavers damage the forest around them, they exert a positive influence overall by regulating water supplies. And while they're at it, they provide habitat for species adapted to large areas of standing water.

So, as we close this chapter, let's return once again to the source of water in the forest—rain. Rain can put you in the most wonderful mood while you are out walking, but if you're not wearing the right clothes, it can be unpleasant. If you live in Europe, mature deciduous trees offer a very special service to help you: a short-term weather forecast brought to you by chaffinches. These rust-red birds with gray heads normally sing a song whose rhythm ornithologists like to transcribe as "chip chip chip chooee chooee cheeoo." But you'll hear that song only on a fine day. If it looks like rain, the song changes to a loud "run run run run run."

19

— YOURS OR MINE? —

THE FOREST ECOSYSTEM is held in a delicate balance. Every being has its niche and its function, which contribute to the well-being of all. Nature is often described like that, or something along those lines; however, that is, unfortunately, false. For out there under the trees, the law of the jungle rules. Every species wants to survive, and each takes from the others what it needs. All are basically ruthless, and the only reason everything doesn't collapse is because there are safeguards against those who demand more than their due. And one final limitation is an organism's own genetics: an organism that is too greedy and takes too much without giving anything in return destroys what it needs for life and dies out. Most species, therefore, have developed innate behaviors that protect the forest from overexploitation. We are already familiar with a good example, and that is the jay that eats acorns and beechnuts but buries a multitude of them as it does so, ensuring that the trees can multiply more efficiently with it than without it.

Whenever you walk through a tall, dark forest, you are walking down the aisles of a huge grocery store. It is filled with all sorts of delicacies—at least as far as animals, fungi, and bacteria are concerned. A single tree contains millions of calories in the form of sugar, cellulose, lignin, and other carbohydrates. It also contains water and valuable minerals. Did I say a grocery store? A better description would be a heavily guarded warehouse, for there is no question here of just helping yourself. The door is barred, the bark thick, and you must come up with a plan to get to the sweet treasures inside. And you are a woodpecker.

Thanks to a unique structure that allows its beak to flex and head muscles that absorb impact, a woodpecker can hack away at trees without getting a headache. In the spring, when water is shooting up through the trees, streaming up to the buds, and delivering delicious provisions, several species of woodpeckers called sapsuckers drill dotted lines of small holes in the thinner trunks or branches. The trees begin to bleed out of these wounds. Tree blood doesn't look very dramatic—it looks a lot like water—however, the loss of this bodily fluid is as detrimental to the trees as it is to us. This fluid is what these sap-sucking woodpeckers are after, and they begin to lick it up. The trees usually mostly tolerate the damage, as long as the woodpeckers don't get carried away and make too many of these holes. Eventually, the holes heal over, leaving patterns that look like intentionally decorative scarring.

Aphids (sometimes also called plant lice or greenflies) are much lazier than woodpeckers. Instead of flying about industriously and hacking out holes here and there, they attach their sucking mouthparts to the veins of leaves and needles. Thus positioned, they get royally drunk in a way no other animals can. The tree's lifeblood rushes right through these tiny insects and comes out the other end in large droplets. Aphids need to saturate themselves like this because the sap contains very little protein—a nutrient they need for growth and reproduction. They filter the fluid for the protein they crave and expel most of the carbohydrates, above all sugar, untouched. Little wonder it rains sticky honeydew under trees infested with aphids. Perhaps you've had the experience of parking your car under a stricken maple only to come back to a thoroughly filthy windscreen.

There are specialized sap-sucking pests for every tree: firs (balsam twig aphid), spruce (green spruce aphid), oaks (oak leaf phylloxera), and beeches (woolly beech aphid). There's sucking and excreting going on everywhere. And because the ecological niche of the leaves is already occupied, there are more species painstakingly boring their way through thick bark to reach the places where sap is flowing underneath. Pests that attack bark, such as woolly beech scale, can completely envelop trunks with their waxy silvery-white wool. These irritants have a similar effect on a tree as scabies has on us: festering wounds appear that take a long time to heal and leave behind rough and scabby bark. Sometimes fungi and bacteria get in and weaken the tree as well, so much so that it dies.

It's no surprise that trees try to defend themselves against these scourges by producing defensive compounds. If the infestations continue, it helps if trees form a thick layer of outer bark to finally get rid of the sap-sucking pests. If they do that, they are protected against further attacks for at least a few years. The possibility of infection is not the only problem. With their voracious appetites, sap-sucking pests remove a gigantic quantity of nutrients from trees. Per square yard of forest, the tiny pests can tap many hundreds of tons of pure sugar from the trees—sugar the trees can no longer use to grow or set aside in reserve for the coming year.

For many animals, however, sap-sucking pests such as aphids are a blessing. First, they benefit other insects such as ladybugs, whose larvae happily devour one aphid after another. Then there are forest ants, which love the honeydew the aphids excrete so much that they slurp it up right from the aphids' backsides. To speed up the process, the ants stroke the aphids with their antennae, stimulating them to excrete the honeydew. And to prevent other opportunists from entertaining the idea of eating the ants' valuable aphid colonies, the ants protect them. There's a regular little livestock operation going on up there in the forest canopy. And whatever the ants can't use doesn't go to waste. Fungi and bacteria quickly colonize the sticky coating that covers the vegetation around the infested tree, and it soon gets covered with black mold. Honeybees also take advantage of aphid excretions. They suck up the sweet droplets, carry them back to their hives, regurgitate them, and turn them into dark forest honey. It is particularly prized by consumers, even though it has absolutely nothing to do with flowers.

Gall midges and wasps are a bit more subtle. Instead of piercing leaves, they reprogram them. To do this, the adults lay their eggs in a beech or oak leaf. The sap-sucking larvae begin to feed, and thanks to chemical compounds in their saliva, the leaf begins to grow into a protective casing or gall. Leaf galls can be pointed (beech) or spherical (oak), but in both cases the young insects inside are protected from predators and can nibble away in peace. When fall comes, the leaf galls fall to the ground together with their occupants, which pupate and then hatch in spring. Particularly in beech trees, there can be massive infestations, but they do very little damage to the tree.

Caterpillars are a different story. What they set their sights on is not sugary sap but leaves and needles in their entirety. If there aren't too many of them, the tree barely notices, but populations explode in regular cycles. I had a run-in with one of these population explosions a few years ago in a stand of oaks in the forest I manage. The trees cover a steep south-facing mountain slope. That June, I noticed with horror that the fresh new leaves had completely disappeared and the trees standing in front of me were as bare as if it were winter. When I got out of my Jeep, I heard a loud roaring like the pounding of rain in a storm. But there was clear blue sky above me, so the noise couldn't be because of the weather. No. It was a hail of feces from millions of oak leaf roller caterpillars. Thousands of black pellets were bouncing off my head and shoulders. Ugh! You can see something similar every year in the large pine forests of eastern and northern Germany. Commercial forest monocultures also encourage the mass reproduction of butterflies and moths, such as nun moths and pine loopers.

What usually happens is that viral illnesses crop up toward the end of the cycle and populations crash.

The caterpillar pellets end in June when the trees are completely stripped of their leaves, and now the trees have to muster their last reserves to leaf out again. Usually, that works just fine. After a few weeks have passed, almost no signs of the feeding frenzy remain; however, tree growth is limited, which you can see in the particularly narrow growth ring in the trunk for that year. If trees are infested and defoliated for two or three years in a row, many of them will weaken and die. Conifer sawflies join the butterfly larvae in the pines. Sawflies "saw" open plant tissue so that they can lay their eggs there. It's not the appetites of the adults but the larvae that the trees have to worry about: up to twelve needles a day disappear into each tiny mouth, which quickly gets dangerous for the tree.

I've already explained, in chapter 2, "The Language of Trees," how trees use scent to summon parasitic wasps and other predators to rid themselves of pests. However, there is yet another strategy they can employ, as demonstrated by the bird cherry. Their leaves contain nectar glands, which secrete the same sweet juice as the flowers. In this case, the nectar is for ants, which spend most of the summer in the trees. And just like people, from time to time these insects crave something heartier than a sugary snack. They get this in the form of caterpillars, and thus they rid the bird cherry of its uninvited guests. But it doesn't always turn out the way the tree intended. The caterpillars get eaten, but apparently, sometimes the amount of sweet nectar the tree provides doesn't satisfy the ants and they begin to farm aphids. As I've explained, these creatures tap into the leaves and when the ants stroke the aphids with their antennae, they exude droplets of sugary liquid for them.

The feared bark beetle basically goes for broke, seeking out weakened trees and trying to move in. Bark beetles live by the principle "all or nothing." Either a single beetle mounts a successful attack and then sends out a scented invitation for hundreds of its kin to come on over and they kill the tree. Or the tree kills the first beetle that bores into it and the buffet is canceled for everyone. The coveted prize is the cambium, the actively growing layer between the bark and the wood. This is where the trunk grows as wood cells form on the inside and bark cells form on the outside. The cambium is succulent and stuffed full of sugar and minerals. In case of emergency, people can also eat it. You can try this out for yourself in the spring. If you come across a spruce recently downed by the wind, cut off the bark with a pocketknife. Then run the blade flat along the exposed trunk and peel off long strips about a third of an inch wide. Cambium tastes like slightly resinous carrots, and it's very nutritious. Bark beetles also find it nutritious—that's why they drill tunnels into the bark so that they can lay their eggs close to this energy source. Well protected from enemies, the larvae can eat here until they are nice and fat.

Healthy spruce defend themselves with terpenes and phenols, which can kill the beetles. If that doesn't work, they dribble out sticky resin to trap them. But researchers in Sweden have discovered that the beetles have been arming themselves. Yet again, the weapons are fungi. These fungi are found on the beetles' bodies. As the beetles make their tunnels, the fungi come along for the ride and end up under the bark. Here, they disarm the spruce's chemical defenses by breaking them down into harmless substances. Because the fungi grow faster than the beetles drill, once they make it under the bark, they are always one step ahead. This means all the terrain the bark beetles encounter has been decontaminated and they can feed safely.[46] Now there is nothing to stop a population explosion, and the thousands of larvae that hatch eventually weaken even healthy trees. Not many spruce can survive such massive attacks.

Large herbivores show less finesse. They need to eat many pounds of food daily, but deep in the forest, food is hard to find. Because there's hardly any light, there isn't much greenery on the forest floor, and the juicy leaves high up in the crown are out of reach. So, in the natural course of things, there aren't many deer in this ecosystem. They get their chance when an old tree falls over. After the fall, light reaches the forest floor for a few years, and not only young trees grow but also, for a short period of time at least, wild flowers and grasses. Animals rush to this oasis of green, which means that any new growth is heavily browsed.

Light means sugar, which makes the young trees attractive to browsers. In the dim light beneath the mother trees, their tiny meager buds usually get hardly any food. What little food they need to survive as they wait for their turn to grow, they get from their parents, who pump it to them via their roots. The sugar-deprived buds are tough and bitter, so the deer pass them by. But as soon as the sun reaches the delicate little trees, they start budding out like mad. Photosynthesis gets underway, the leaves get thicker and juicier, and the buds, which form over the summer to break out the next spring, are all over the youngsters and full of nutrients. And that's the way it should be, because the next generation of trees wants to step on it and grow upward as quickly as it can before the window of light closes again. But all this activity attracts the deer's attention, and they don't want to miss out on the delicacies on offer. And now the competition between the young trees and the deer heats up for a few years. Will the little beeches, oaks, and firs manage to grow tall enough fast enough so that the animals can no longer get their mouths around the all-important main shoots? Usually, the deer don't destroy all the little trees in one small group, so there are always a couple that escape damage and battle on upward. Those whose leading shoots have been nibbled off now grow bowed and bent, and they are soon overtaken by undamaged shoots. Eventually, damaged youngsters will die from light deprivation and return to humus.

One rascal that does more damage than it looks as though it should from its size is the honey fungus mushroom, that innocuous-looking fruiting body that often appears on tree stumps in the fall. They are found in both Europe and North America. None of the seven honey fungus mushrooms native to Central Europe (it's difficult to tell them apart) do trees any good. Quite the opposite, in fact. Their mycelium—white underground threads—force their way into the roots of firs, beeches, oaks, and other species of tree. Eventually, they grow up under the bark, where they form white fan-shaped patterns. The bounty they steal—at first mostly sugar and nutrients out of the cambium—is carted off in what look like thick black cords. These rootlike structures, which are sometimes referred to as "boot laces" because of their appearance, are unique in the world of fungi. But honey fungus doesn't content itself with sugary treats. As it continues to develop, it eats through wood as well and causes its host tree to rot. At the end of the process, the tree eventually dies.

Pinesap, which belongs to the same family as blueberries and heathers, is much more subtle. It doesn't contain any green pigment and manages to grow only a nondescript light-brown flower. A plant that isn't green doesn't contain any chlorophyll and, therefore, cannot photosynthesize. This means the pinesap depends on others for food. It insinuates itself with mycorrhizal fungi—the ones helping the trees' roots—and because it doesn't photosynthesize and therefore doesn't need any light, it can grow in even the darkest stands of spruce. There it taps into the flow of nutrients traveling between the fungi and the tree roots, siphoning off a portion for itself. Small cow wheat does something similar, only rather more sanctimoniously. It also loves spruce and also hooks up to the root-fungi system, joining the feast as an uninvited guest. Although its aboveground parts are typical plant green and can indeed turn a bit of light and carbon dioxide into sugar, they are mostly a display to disguise what's really going on.

Trees offer considerably more than just food. Animals abuse young trees by using them as convenient rubbing posts.

For instance, every year, male deer have to get rid of the skin or "velvet" from their soon-to-be-shed antlers. So they search out a little tree that is sturdy enough not to break easily and, at the same time, also slightly flexible. Here, the lords of creation let loose for days on end until the last scrap of itchy skin has been rubbed off. The little tree's bark is in such a bad state after this performance that the tree usually dies. When they are choosing their trees, deer go for whatever is unusual. Whether they choose spruce, beech, pine, or oak, they will always choose whatever is uncommon locally. Who knows? Perhaps the smell of the shredded bark acts like an exotic perfume. It's the same with people: it's the rare things that are most highly prized.

Once the diameter of the trunk exceeds 4 inches, it's game over. By then, the bark of most species of trees is so thick that it can withstand the impetuous antlered beasts. In addition, the trunks are now so stable that they no longer bend, and they are too wide to fit between the tips of the deer's antlers. But now the deer need something else. Normally, they wouldn't be living in the forest at all because they eat mostly grass. Grass is a rarity in a natural forest and almost never present in the quantities a whole herd requires, and therefore, these majestic animals prefer to live out in the open. But river valleys, where floods ensure open grassland, are where people like to live. Every square yard is used for urban areas or agriculture. And so the deer have retreated to the forest, even though they sneak out at night. But as typical plant eaters, they need fiber-rich food around the clock. When there isn't anything else, in desperation, they turn to tree bark.

When a tree is full of water in the summer, it's easy to peel off its bark. The deer bite into it with their incisors (which they have only on their lower jaws) and pull off whole strips from the bottom up. In winter, when the trees are sleeping and the bark is dry, all the deer can do is tear off chunks. As always, this activity is not only really painful for trees but also life threatening. There's often a large-scale fungal invasion through the huge gaping wounds, which quickly breaks down the wood. The damage is so extensive the tree can't close the wound by quickly walling it off. If the tree grew up in an undisturbed forest—that is to say, nice and slowly—it can survive even severe setbacks like this. Its wood is made up of the tiniest rings, so it's tough and dense, which makes things very difficult for the fungi that are trying to work their way into it. I have often seen tree youngsters like this that have managed to close wounds, even though it took them decades. It's quite another story with the planted trees in our commercial forests. Usually, they grow very quickly and their growth rings are huge; therefore, their wood contains a great deal of air. Air and moisture—these are ideal conditions for fungi. And so the inevitable happens: severely damaged trees snap in middle age. If, however, the wounds inflicted over the winter are numerous but small, the tree can close them up without suffering any long-term damage.

20

— COMMUNITY HOUSING — PROJECTS

EVEN IF MATURE trees have now grown too thick for many of the activities I have described so far, animals are happy to go on using them. These giants can become coveted living spaces, a service the trees do not offer voluntarily. Birds, martens, and bats are particularly partial to the thick trunks of older trees. They like thick trunks because the sturdy walls provide especially good insulation against heat and cold.

In Europe, it's usually a great spotted woodpecker or a black woodpecker that gets things started. The bird hacks out a hole in the trunk that may be only an inch or two deep. Contrary to popular opinion, the birds don't restrict themselves to rotten trees, and they often start construction in healthy trees. Would you move into a ramshackle home if you could build a new one next door? Just like us, woodpeckers want the place where they bring up their families to be solid and durable. Even though the birds are well equipped to hammer away at healthy wood, it would be too much for them to complete the job all at once. And that's why they take a months-long break after the first phase, hoping fungi will pitch in. As far as the fungi are concerned, this is the invitation they have been waiting for, because usually they can't get past the bark. In this case, they quickly move into the opening and begin to break down the wood. What the tree sees as a coordinated attack, the woodpecker sees as a division of labor. After a while, the wood fibers are so mushy that it's much easier for the woodpecker to enlarge the hole.

Finally, the day comes when construction is complete and the cavity is ready to move into. But that's not enough for the crow-sized black woodpecker, and so he works on a number of cavities at the same time. He uses one for the kids, one for sleeping, and the others for a change of scene. Every year, the cavities are renovated, and wood chips at the base of the trees are evidence of this activity. Renovation is necessary because the fungi that have invaded the space are by now unstoppable. They keep eating deeper into the trunk, transforming the wood into damp mush, which isn't an ideal environment in which to raise a family. Every time the woodpecker cleans out the soggy mess, the nesting cavity gets a little larger. Sooner or later, the cavity becomes too large and, above all, too deep for the baby birds, which must climb up out of the opening to make their first flight. And now it's time for the subletters to move in.

The subletters are species that can't work with wood themselves. There's the nuthatch, which is somewhat like a woodpecker but much smaller. Like woodpeckers, it hops around on dead wood, pecking away to get at beetle larvae. It loves to build its nests in abandoned great spotted woodpecker nesting cavities. But there's a problem. The entrance is way larger than it needs and could let in predators intent on eating its brood. To prevent this, the bird makes the entrance smaller using mud, which it arranges artfully around the perimeter.

While we're on the subject of predators: trees also offer their subletters a special service on the side, thanks to the characteristics of their wood. Wood fibers conduct sound particularly well, which is why they are used to make musical instruments such as violins and guitars. You can do a simple experiment to test for yourself how well these acoustics work. Put your ear up against the narrow end of a long trunk lying on the forest floor and ask another person at the thicker end to carefully make a small knocking or scratching sound with a pebble. On a still day, you can hear the sound through the trunk incredibly clearly, even if you lift your head. Birds use this property of wood as an alarm system for their nesting cavities. In their case, what they pick up is not benign knocking but scrabbling sounds made by the claws of martens or squirrels. The sounds can be heard high up in the tree, which gives the birds a chance to escape. If there are young in the nest, they can try to distract the attackers, though such attempts are usually doomed to failure. But at least the parents escape with their lives and can compensate for their loss by raising a second brood.

Acoustics are not so important for bats, for they have completely different concerns. Some species of these tiny mammals need lots of tree cavities at the same time to raise their young. In the case of Bechstein's bats, which live in Europe and western Asia, small groups of females raise their offspring together. They spend only a few days in the same quarters before it's time to move on. The reason for this is parasites. If the bats were to spend the whole season in the same cavity, there would be a parasite population explosion and they would torment the winged nocturnal hunters mercilessly. Short moving cycles take care of this by simply leaving the parasites behind.

Owls don't fit very well into woodpecker cavities, and so they must be patient for a few more years. Over time, the tree continues to rot, and sometimes the trunk splits open a bit more so that the entrance gets bigger. And sometimes there is a series of woodpecker cavities up the trunk that speeds the owls' entrance. These are like woodpecker apartments stacked one on top of the other. As the process of decay progresses, they slowly merge into each other, and when that happens, they are ripe for the arrival of the tawny owl and his friends.

And what about the tree? Its efforts to defend itself are in vain. And it's too late to mount an attack against fungi anyway, because by now the floodgates have been open to them for years. But the tree can lengthen its life-span considerably if it at least manages to get a grip on its external wounds. If it manages to do this, it will continue to rot on the inside; however, externally it will be as stable as a hollow steel pipe and can survive for another hundred years. You can spot these attempts at repair if you see bulges around the edges of a woodpecker hole. Despite its best efforts, the tree rarely makes headway on closing the entrance. Usually, the merciless builder simply pecks the new wood away.

The rotting trunk now becomes home to a complex living community. Wood ants move in and chew the moldy wood to make their papery nests. They soak the nest walls with honeydew, the sugary excretions of aphids. Fungi bloom on this substrate, and their fibrous web stabilizes the nest. A multitude of beetles are drawn to the mushy, rotten interior of the cavity. Their larvae can take years to develop, and therefore, they need stable, long-term accommodations. This is why they choose trees, which take decades to die and, therefore, remain intact for a long time. The presence of beetle larvae ensures that the cavity remains attractive to fungi and other insects, which keep a constant supply of excrement and sawdust raining down into the rot.

The excrement of bats, owls, and dormice also drops down into the dark depths. And so the rotten wood is constantly supplied with nutrients, which feed species such as the blood-necked click beetle,[47] or the larvae of the European hermit beetle, a big black beetle that can grow up to 1.5 inches long. Hermit beetles are very reluctant to move and prefer to spend their whole lives in a dark hole at the base of a rotting tree trunk. And because these beetles rarely fly or walk, whole generations of the same family can live for decades in the same tree. And this explains why it is so important to keep old trees. If they are cleared away, these little black guys can't just wander over to the next tree; they simply don't have the energy to do that.

Even if one day the tree gives up and breaks off in a storm, it has still served the community well. Even though scientists haven't fully researched the relationships yet, we do know that higher species diversity stabilizes the forest ecosystem. The more species there are around, the less chance there is that a single one will take over to the detriment of the others, because there's always a candidate on hand to counteract the menace. And even the dead tree trunk can offer a valuable service managing water for living trees merely because it is there, as we've already seen in chapter 17, "Woody Climate Control."

21

MOTHER SHIPS OF BIODIVERSITY

MOST ANIMALS THAT depend on trees don't harm them. They just use the trunks or the crowns as custom-built homes that offer small ecological niches, thanks to varying amounts of moisture and light. Innumerable specialists find places to live here. Little research has been done, particularly in the upper levels of the forest, because scientists need to use expensive cranes or scaffolds to check them out. To keep costs down, brutal methods are sometimes employed. And so, in 2009, tree researcher Dr. Martin Gossner sprayed the oldest (six hundred years old) and mightiest (170 feet tall and 6 feet wide at chest height) tree in the Bavarian Forest National Park. The chemical he used, pyrethrum, is an insecticide, which brought any number of spiders and insects tumbling down to the forest floor—dead.

The lethal results show how species-rich life is way up high. The scientist counted 2,041 animals belonging to 257 different species.[48]

Tree crowns even contain specialized wetland habitats. When a trunk splits to form a fork, rainwater collects at the point where the trunk divides. This minuscule pool is home to tiny little flies that provide food for rare species of beetles. It's more difficult for animals to live in trunk cavities where water collects. The cavities are dark, and the murky, moldy brew that lurks there contains very little oxygen. Larvae that develop in water cannot breathe in places like that—unless, of course, they are endowed with snorkels, like the offspring of the bumblebee hoverfly. Thanks to breathing tubes that extend like telescopes, these larvae can survive here. Bacteria are almost the only things stirring in these waters, so they are probably the larvae's food source.[49]

Not every tree is targeted by woodpeckers as a nesting site and doomed to gradual rot, and by no means do all slowly waste away, offering many specialized species hard-to-find habitats as they do so. Many trees die quickly. A storm might snap a mighty trunk, or bark beetles might destroy a tree's bark in a few short weeks, causing its leaves to wither and die. Then the ecosystem around the tree changes suddenly. Animals and fungi that are dependent on the tree pumping a steady supply of moisture through its veins or sugar to its crown must now leave the corpse or starve. A small world has come to an end. Or has it just begun?

"*Und wenn ich geh, dann geht nur ein Teil von mir.*" "And when I go, only a part of me is gone." This phrase from a hit by

German pop singer Peter Maffay could have been written by a tree. For the dead trunk is as indispensable for the cycle of life in the forest as the live tree. For centuries, the tree sucked nutrients from the ground and stored them in its wood and bark. And now it is a precious resource for its children. But they don't have direct access to the delicacies contained in their dead parents. To access them, the youngsters need the help of other organisms. As soon as the snapped trunk hits the ground, the tree and its root system become the site of a culinary relay race for thousands of species of fungi and insects. Each is specialized for a particular stage of the decomposition process and for a particular part of the tree. And this is why these species can never pose a danger to a living tree—it would be much too fresh for them. Soft, woody fibers and moist, moldy cells—these are the things they find delicious. They take their sweet time over both their meals and their life cycles, as demonstrated by the stag beetle. The adult beetle lives for only a few weeks, just long enough to mate. This animal spends most of its life as a larva, which slowly eats its way through the crumbling roots of dead deciduous trees. It can take up to eight years for it to get big and fat enough to pupate.

Bracket fungus is similarly slow. It gets its name because it sticks out from the dead trunk like a shelf made from half a broken plate. The red belt conk is one example. It feeds on the white threads of cellulose in the wood, leaving brown crumbs as evidence of its meal. Its fruiting body, the aforementioned broken plate, is attached to the trunk at a neat horizontal angle. This is the only way it can ensure that its reproductive spores will trickle out of the small tubes on its underside. If the rotten tree it is attached to falls over one day, the fungus seals the tubes and continues to grow at right angles to its former fruiting body so that it can form a new horizontal plate.

Some fungi fight bitterly over feeding territory. You can see this clearly on dead wood that has been sawn into pieces. You'll find marbled structures of lighter and darker tissue clearly separated by black lines. The different shades indicate different species of fungus working their way through the wood. They wall off their territory from other species with dark, impenetrable polymers, which look to us as though they are drawing battle lines.

In total, a fifth of all animal and plant species—that's about six thousand of the species we know about—depend on dead wood.[50] As I have explained, dead wood is useful because of its role as a nutrient recycler. But can it also be a threat to the forest? After all, perhaps if there's not enough dead wood lying around, organisms might decide to eat live trees instead. I hear this concern voiced time and again by people who come to visit the forest, and there are a few private forest owners who remove all dead trunks for exactly this reason. But this is neither a necessary nor a useful practice. All removing dead wood does is destroy valuable habitats, because live wood is of no use to organisms that live in dead wood. Live wood is not soft enough for them, it is too moist, and it contains too much sugar. This is quite apart from the fact that beeches, oaks, and spruce defend themselves from colonization. Healthy trees growing in their natural range withstand almost all attacks if they are well nourished. And the armada of decomposers helps feed the living trees as long as the little guys can find a way to make their livelihoods.

Sometimes dead wood is directly beneficial to trees, for example, when a downed trunk serves as a cradle for its own young. Young spruce sprout particularly well in the dead bodies of their parents. This is known as "nurse-log reproduction" in English and, somewhat gruesomely, as *Kadaververjüngung*, or "cadaver rejuvenation," in German. The soft, rotten wood stores water particularly well, and some of the nutrients it contains have already been released by fungi and insects. There is just one teeny problem: the trunk isn't a permanent replacement for soil, as it is constantly being degraded, until one day it disintegrates completely into humus on the forest floor. So what happens to the young trees then? Their roots are exposed and lose their support, but because the process plays out over decades, the roots follow the disintegrating wood into the forest floor. The trunks of spruce that grow up this way end up being elevated on stilts. The height of the stilts corresponds to the diameter of the nurse log on which they once lay.

22

— HIBERNATION —

It's late summer, and the forest is in a strange mood. The trees have exchanged the lush green in their crowns for a washed-out version verging on yellow. It seems as though they are getting increasingly tired. Exhaustion is setting in, and the trees are waiting for the stressful season to end. They feel just like we do after a busy day at work—ready for a well-earned rest.

Grizzly bears hibernate and so do dormice. But trees? Do they experience anything that could be compared to our nightly time-outs? The grizzly bear is a good candidate for comparison, because it follows a similar strategy to trees. In summer and early fall, it eats to lay down a thick layer of fat it can live off all winter. And this is exactly what our trees do as well. Of course, they don't feed on blueberries or salmon, but they fuel themselves with energy from the sun, which they use to make sugar and other compounds they can hold in reserve. And they store these under their skin just like a bear. Because they can't get any fatter (only their bones—that is to say, their wood—can grow), the best they can do is fill their tissues with food. And whereas bears can go on eating everything they can find, at some point, the trees get full.

You can see this very well especially if you look at wild cherries, bird cherries, and wild service trees any time after August. Even though there are many beautiful sunny days they could make use of before October, they begin to turn red. And what that means is that they are shutting up shop for the year. The storage spaces under their bark and in their roots are full. If they made more sugar, there would be nowhere to stash it. While the bears happily go on eating, for these trees the sandman is already knocking on the door. Most other tree species seem to have larger storage areas, and they continue to photosynthesize hungrily and without taking a break right until the first hard frosts. Then they, too, must stop and shut down all activity. One reason for this is water. It must be liquid for the tree to work with it. If a tree's "blood" freezes, not only does nothing work, but things can also go badly wrong. If wood is too wet when it freezes, it can burst like a frozen water pipe. This is the reason most species begin to gradually reduce the moisture content in their wood—and this means cutting back on activity—as early as July.

But trees can't switch to winter mode yet, for two main reasons. First, unless they are members of the cherry family, they use the last warm days of late summer to store energy, and second, most species still need to fetch energy reserves from the leaves and get them back into their trunk and roots. Above all, they need to break down their green coloring, chlorophyll, into its component parts so that the following spring they can send large quantities of it back out to the new leaves. As this pigment is pumped out of the leaves, the yellow and brown colors that were there all along predominate. These colors are made of carotene and probably serve as alarm signals. Around this time, aphids and other insects are seeking shelter in cracks in the bark, where they will be protected from low temperatures. Healthy trees advertise their readiness to defend themselves in the coming spring by displaying brightly colored fall leaves.[51] Aphids & Co. recognize these trees as unfavorable places for their offspring because they will probably be particularly vigorous about producing toxins. Therefore, they search out weaker, less colorful trees.

But why bother with all this extravagance? Many conifers demonstrate that things can be done differently. They simply keep all their green finery on their branches and thumb their noses at the idea of an annual makeover. To protect its needles from freezing, a conifer fills them with antifreeze. To ensure it doesn't lose water to transpiration over the winter, it covers the exterior of its needles with a thick layer of wax. As an extra precaution, the skin on its needles is tough and hard, and the small breathing holes on the underside are buried extra deep. All these precautions combine to prevent the tree from losing any significant amount of water. Such a loss would be tragic, because the tree wouldn't be able to replenish supplies from the frozen ground. It would dry out and could then die of thirst.

In contrast to needles, leaves are soft and delicate—in other words, they are almost defenseless. It's little wonder beeches and oaks drop them as quickly as they can at the first hint of frost. But why didn't these trees simply develop thicker skins and antifreeze over the course of their evolution? Does it really make sense to grow millions of new leaves per tree every year, use them for a few months, and then go to the trouble of discarding them again? Apparently, evolution says it does, because when it developed deciduous trees about 100 million years ago, conifers had already been around on this planet for 170 million years. This means deciduous trees are a relatively modern invention. When you take a closer look, their behavior in fall actually makes a lot of sense. By discarding their leaves, they avoid a critical force—winter storms.

When storms blow through forests in Central Europe from October on, it's a matter of life and death for many trees. Winds blowing at more than 60 miles an hour can uproot large trees, and some years, 60 miles an hour is a weekly occurrence. Fall rains soften the forest floor, so it's difficult for tree roots to find purchase in the muddy soil. The storms pummel mature trunks with forces equivalent to a weight of approximately 220 tons. Any tree unprepared for the onslaught can't withstand the pressure and falls over. But deciduous trees are well prepared. To be more aerodynamic, they cast off all their solar panels. And so a huge surface area of 1,200 square yards disappears and sinks to the forest floor.[52] This is the equivalent of a sailboat with a 130-foot-tall mast dropping a 100-by-130-foot mainsail. And that's not all. The trunk and branches are shaped so that their combined wind resistance is somewhat less than that of a modern car. Moreover, the whole construction is so flexible that the forces of a

strong gust of wind are absorbed and distributed throughout the tree.

These measures all work together to ensure that hardly anything happens to deciduous trees over the winter. If there's an unusually strong hurricane-force wind—the kind that happens only every five to ten years in Europe—the tree community stands together to help each individual tree. Every trunk is different. Each has its own pattern of woody fibers, a testament to its unique history. This means that, after the first gust—which bends all the trees in the same direction at the same time—each tree springs back at a different speed. And usually it is the subsequent gusts that do a tree in, because they catch the tree while it's still severely bowed and bend it over again, even farther this time. But in an intact forest, every tree gets help. As the crowns swing back up, they hit each other, because each of them is straightening up at its own pace. While some are still moving backwards, others are already swinging forward again. The result is a gentle impact, which slows both trees down. By the time the next gust of wind comes along, the trees have almost stopped moving altogether and the struggle begins all over again. I never tire of watching tree crowns move back and forth. I can see both the movement of the whole community and the movements of individual trees. Bear in mind, however, that it's never a good idea to go into the forest during a storm.

Let's get back to the subject of dropping leaves. With every winter they survive, the trees prove that this makes sense and that producing new leaves every year is worth the energy it takes. But it brings up completely different dangers. One of these is snowfall. Snow makes it imperative that deciduous trees drop their leaves in a timely manner. Once the aforementioned 1,200 square yards of leaf surface have disappeared, the white blanket has no place to land but on the branches, and this means that most of it falls through onto the ground.

Ice can generate even heavier loads than snow. A few years ago, I experienced weather conditions that combined temperatures slightly below freezing with a seemingly harmless drizzle. This unusual weather lasted for three days, and as each hour passed, I became more and more worried about the forest. The light rain landing on freezing branches turned to ice in seconds, quickly weighing the branches down. It looked incredibly beautiful. All the trees were encased in crystal. Whole stands of young birches were bent down under the weight of the ice, and with a heavy heart, I was already giving them up for lost. In the case of mature trees, it was above all the conifers—mostly Douglas firs and pines—that lost up to two thirds of the green branches in their crowns, which broke off with a loud cracking sound. That weakened the trees considerably, and it will take decades for them to completely rebuild their crowns. But the bent-over young birches surprised me. When the ice melted several days later, 95 percent of the trunks stood tall again. Today, a few years later, there's no sign that anything happened to the trees. Of course, there were a few that didn't manage to spring back. They died, at some point their rotten little trunks broke, and they are now slowly turning themselves into humus.

So, dropping leaves is an effective protective strategy that seems made to measure for the climate in Central European latitudes. It is also an opportunity for trees to finally excrete waste. Just as we take a trip to a quiet little room before we go to bed, trees also rid themselves of substances they do not need and would like to part with. These drift down to the ground in their discarded leaves. Shedding leaves is an active process, so the tree can't go to sleep yet. After the reserve supplies have been reabsorbed from the leaves back into trunk, the tree grows a layer of cells that closes off the connection between the leaves and the branches. Now all it takes is a light breeze, and the leaves drift down to the ground. Only when that process is complete can trees retire to rest. And this they must do to recuperate from the exertions of the previous season. Sleep deprivation affects trees and people in much the same way: it is life threatening. That's why oaks and beeches can't survive if we try to grow them in containers in our living rooms. We don't allow them to get any rest there, and so most of them die within the first year.

Young trees standing in their parents' shadow exhibit a few clear deviations from the standard strategy for shedding leaves. When the mother trees lose their leaves, sunlight suddenly floods the ground. The eager young pups are waiting for just this moment, and they take advantage of the bright light to fill up with lots of energy—and they are usually surprised by the first frosts while they are at it. If temperatures are well below freezing, with nights lower than 23 degrees Fahrenheit, the trees have no option but to start yawning and begin hibernation. Now it's too late to grow a separating layer of cells, and jettisoning leaves is no longer an option; however, this is no big deal for the tiny trees. Because they are so small, the wind is no threat and even snow is rarely a problem.

In the spring, the young trees exploit a similar opportunity. They leaf out two weeks before the large trees, ensuring themselves a long leisurely breakfast in the sun. But how do the youngsters know when they need to get started? After all, they don't know the date when the mother trees might leaf out. It's warm temperatures close to the ground that give the game away. Spring really is rung in here approximately two weeks earlier than it is 100 feet higher up in the canopy. Up high, harsh winds and freezing cold nights delay the warm season for a little while longer. It's the protective canopy created by the branches of the old trees that keeps heavy, late frosts from reaching the ground. At the same time, the layer of leaves covering the soil acts like a warming compost pile, allowing the thermometer to climb a couple of degrees. Counting the days they benefited from in the fall, the youngsters can enjoy one month of free growth time—and that's almost 20 percent of the growing days available to them. Not bad.

Among deciduous trees, there are different approaches to frugal living. Most trees draw energy reserves back into their branches before they shed their leaves, but a few don't seem to care. Alders, for example, happily drop bright-green leaves onto the ground as though there were no tomorrow. Alders, however, usually grow in swampy, nutrient-rich soil and can, apparently, afford the luxury of producing new chlorophyll every year. Fungi and bacteria at the base of the trees recycle the discarded leaves to produce the raw materials the alders need to build chlorophyll, and all the trees need to do is take these building blocks up through their roots. They don't even have to worry about recycling nitrogen, thanks to the symbiotic relationship they have with bacteria in nodules on their roots, which constantly provide them with all the nitrogen they need. Per year and square mile of alder forest, these tiny helpers can extract up to 87 tons of nitrogen from the air and make it available to the roots of their tree friends.[53] That is more than most farmers spread over their fields as fertilizer.

So, whereas many trees take pains to budget carefully, alders flaunt their wealth. Ash and elders behave in a similar manner. Because these spendthrifts all discard their leaves while they are still green, they don't contribute anything to the fall colors of the forest. Only the misers, it seems, are colorful. No, that's not quite true. Yellow, orange, and red come to the fore when chlorophyll is removed, but these carotenes and anthocyanins are also broken down eventually. The oak is such a careful species that it stashes everything away and discards only brown leaves. Thus, trees differ in their spending habits. It's all over for the beech when its leaves turn brown and yellow, whereas the cheery cherry loses its leaves when they're red.

Finally, we return to the conifers. I've given them rather short shrift so far, but there are three species that drop their leaves like deciduous trees—the larch, the bald cypress, and the dawn redwood. I have no idea why these three conifers are the only ones to follow the deciduous trees' example. Perhaps in the evolutionary competition the best way to overwinter has simply not yet been decided. Holding on to needles certainly brings advantages in the spring, because the trees can get going immediately without having to wait for new growth. However, many new shoots dry out when the crowns warm up nicely in the spring sun and begin to photosynthesize while the ground is still frozen. Because they can't put the brakes on transpiration, as soon as they become aware of the danger, the needles go limp—particularly those from last year, which don't yet have a thick coat of wax.

Apart from that, spruce, pines, firs, and Douglas firs change out their needles because they too must rid themselves of waste materials. They shed the oldest needles, which are damaged and don't work very well anymore. As long as the trees are healthy, firs always keep ten, spruce six, and pines three years' worth of needles, as you can tell by taking a look at the annual growth intervals on their branches. Pines especially, which shed about a quarter of their green needles, can look somewhat sparse in winter. In spring, a new year's worth of needles is added along with fresh growth, and the crowns look the picture of health once again.

ASPEN

23

— A SENSE OF TIME —

IN MANY LATITUDES, forests drop leaves in the fall and leaf out in the spring, and we take this cycle for granted. But if we take a closer look, the whole thing is a big mystery, because it means that trees need something very important: a sense of time. How do they know that winter is coming or that rising temperatures aren't just a brief interlude but an announcement that spring has arrived?

It seems logical that warmer days trigger leaf growth, because this is when frozen water in the tree trunk thaws to flow once again. What is unexpected is that the colder the preceding winter, the earlier the leaves unfurl. Researchers from the Technical University of Munich (TUM) tested this in a climate-controlled laboratory.[54] The warmer the cold season, the later beech branches greened up—and at first glance, that doesn't seem logical. After all, in warm years, lots of other plants—wild flowers, for example—often start to grow in January and even begin to flower, as we are constantly reminded by media headlines. Perhaps trees need freezing temperatures to get a restorative sleep in winter and that's why they don't get going right away in the spring. Whatever the reason, in these times of climate change, this is a disadvantage, because other species that are not so tired and grow their new leaves more quickly will be a step ahead.

How often have we experienced warm spells in January or February without the oaks and beeches greening up? How do they know that it isn't yet time to start growing again? We've begun to solve the puzzle with fruit trees, at least. It seems the trees can count! They wait until a certain number of warm days have passed, and only then do they trust that all is well and classify the warm phase as spring.[55] But warm days alone do not mean spring has arrived.

Shedding leaves and growing new ones depends not only on temperature but also on how long the days are. Beeches, for example, don't start growing until it is light for at least thirteen hours a day. That in itself is astounding, because to do this, trees must have some kind of ability to see. It makes sense to look for this ability in the leaves. After all, they come with a kind of solar cell, which makes them well equipped to receive light waves. And this is just what they do in the summer months, but in April the leaves are not yet out. We don't yet understand the process completely, but it is probably the buds that are equipped with this ability. The folded leaves are resting peacefully in the buds, which are covered with brown scales to prevent them from drying out. Take a closer look at these scales when the leaves start to grow and hold them up to the light. Then you'll see it. They're transparent! It probably takes only the tiniest amount of light for the buds to register day length, as we already know from the seeds of some agricultural weeds. Out in the fields, all it takes is the weak light of the moon at night to trigger germination. And a tree trunk can register light as well. Most tree species have tiny dormant buds nestled in their bark. When a neighboring tree dies and falls down, more sun gets in, which in many trees triggers the growth of these buds so that the tree can take advantage of the additional light.

And how do trees register that the warmer days are because of spring and not late summer? The appropriate reaction is triggered by a combination of day length and temperature. Rising temperatures mean it's spring. Falling temperatures mean it's fall. Trees are aware of that as well. And that's why species such as oaks or beeches, which are native to the Northern Hemisphere, adapt to reversed cycles in the Southern Hemisphere if they are exported to New Zealand and planted there. And what this proves as well, by the way, is that trees must have a memory. How else could they inwardly compare day lengths or count warm days?

In particularly warm years, with high fall temperatures, you can find trees whose sense of time has become confused. Their buds swell in September, and a few trees even put out new leaves. Trees that get in a muddle like this have to suffer the consequences when delayed frosts finally arrive. The fresh growth has not had time to get woody—that is, to get hard and tough for winter—and the leaves are defenseless anyway. And so the new greenery freezes, and that must surely hurt. Worse, the buds for next spring are now lost and costly replacements must be grown. If a tree isn't careful, it will deplete its energy supplies and be less prepared for the coming season.

Trees need a sense of time for more than just their foliage. This sense is equally important for procreation. If their seeds fall to the ground in fall, they mustn't sprout right away. If they do, two problems present themselves. First, the delicate shoots won't have time to get woody, which means they will freeze. Second, when the weather is cold, there is very little for deer to eat and they would be only too happy to pounce on the fresh, green growth. So it's better to sprout in the spring along with all the other plants. Therefore, seeds register cold, and only when extended warm periods follow hard frost do the baby trees dare to come out of their protective coverings. Many seeds don't possess a sophisticated counting mechanism like the one used to trigger leaf growth, and that's why it works so well when squirrels and jays bury beechnuts and acorns an inch or so deep in the soil. Down here it doesn't warm up until true spring arrives. Light seeds, such as the seeds of birches, have to pay more attention. With their little wings, they always land on the surface of the soil and just lie there. Depending on where they come to rest, they might end up in bright sunlight, and therefore, these little ones must be able to wait and register the appropriate day length just as their parents do.

24

— A QUESTION OF CHARACTER —

ON THE COUNTRY road between my home village of Hümmel and the next small town in the Ahr valley stand three oaks. They are a commanding presence out in the open fields, and the area is named in their honor. They are growing unusually close together: mere inches separate the one-hundred-year-old trunks. That makes them ideal subjects for me to study, because the environmental conditions for all three are identical. Soil, water, local microclimate—there can't be three different sets of each within a few yards. This means that if the oaks behave differently, it must be because of their own innate characteristics. And they do, indeed, behave differently!

In winter, when the trees are bare, or in summer, when they are in full leaf, the driver of a car speeding by wouldn't even notice three separate trees. Their interconnecting crowns form a single large dome. The closely spaced trunks could all be growing from the same root, as happens sometimes if downed trees start to regrow. However, the triad of fall color points to a very different story. Whereas the oak on the right is already turning color, the middle one and the one on the left are still completely green. It takes a couple of weeks for the two laggards to follow their colleague into hibernation. But if their growing conditions are identical, what accounts for the differences in their behavior? The timing of leaf drop, it seems, really is a question of character.

As we learned in previous chapters, a deciduous tree has to shed its leaves. But when is the optimal moment? Trees cannot anticipate the coming winter. They don't know whether it is going to be harsh or mild. All they register are shortening days and falling temperatures. If temperatures are falling, that is. There are often unseasonably warm days in the fall, and now the three oaks find themselves in a dilemma. Should they use these mild days to photosynthesize a while longer and quickly stash away a few extra calories of sugar? Or should they play it safe and drop their leaves in case there's a sudden frost that forces them into hibernation? Clearly, each of the three trees decides differently.

The tree on the right is a bit more anxious than the others, or to put it more positively, more sensible. What good are extra provisions if you can't shed your leaves and have to spend the whole winter in mortal danger? So, get rid of the lot in a timely manner and move on to dreamland! The two other oaks are somewhat bolder. Who knows what next spring will bring, or how much energy a sudden insect attack might consume and what reserves will be left over afterward? Therefore, they simply stay green longer and fill the storage tanks under their bark and in their roots to the brim. Until now, this behavior has always paid off for them, but who knows how long it will continue to do so? Thanks to climate change, fall temperatures are remaining high for longer and longer, and the gamble of holding on to leaves is being drawn out until November. All the while, fall storms are beginning as punctually as ever in October, and so the risk of getting blown over while still in full leaf rises. In my estimation, more cautious trees will have a better chance of surviving in the future.

You can see something similar on the trunks of deciduous trees and silver firs. According to the tree etiquette manual, the trunks should be tall and smooth, and this means no branches on the lower half of the tree. That makes sense because there's not much light at the bottom. As there are no sunbeams to be processed, unnecessary body parts that would only use up food are simply shut down. It's a bit like our muscles, which our bodies reduce in size when we don't use them in order to save calories. But trees cannot remove their branches on their own; they just have to let them die. The rest must be done by fungi, which attack the wood once it is dead. At some point, the branch rots, breaks off, and is finally recycled into humus.

Now the tree has a problem at the point where the branch broke off. Fungi can easily grow farther into the trunk because there is no protective coating of bark—at least not yet. But the tree can change this. If the branches were not too thick (up to an inch across), it takes just a few years for the tree to close the gap. The tree can then saturate the area with water from the inside, killing the fungi. But if the branches were very thick, this procedure takes too long. The wounds gape open for decades, offering portals through which the fungi can enter and penetrate deep into the wood. The trunk rots and, at the very least, becomes a little less stable. And that is precisely the reason the etiquette manual calls for only thin branches on the lower part of the trunk. Once they have fallen off as the tree grows, under no circumstances are they to be replaced. Yet that is exactly what a few trees do.

When a neighboring colleague dies, some trees use the light that falls on them to grow out new buds below. They grow thick branches that are very beneficial at first. These trees can now take advantage of the opportunity to photosynthesize in two places at once: at the crown and lower down on the trunk. But one day, perhaps twenty years later, the other trees standing around will have increased the size of their crowns so much that the gap in the canopy closes up. Once again, the lower levels are dark, and the thick branches die. Now the trees pay dearly for their craving for sun. As I've just described, fungi now march deep into the trunk of the foolish trees and put them in danger. When you take your next walk into the forest, you can check for yourself to see that such behavior really is an individual choice and, therefore, a question of character. Take a look at the trees growing around a small clearing. All have the same temptation to do something stupid like growing new branches on their trunks, but only a few give in. The rest keep their bark nice and smooth and avoid the predictable risk.

25

— THE SICK TREE —

STATISTICALLY SPEAKING, MOST species of trees can live to a ripe old age. In the burial area of the forest I manage, tree buyers always ask how long their tree might live. Mostly, they choose beeches or oaks, and as far as we know, these trees usually live to be between four hundred and five hundred years old. But what is a statistic worth when you apply it to an individual tree? Just as much as it is worth when you apply it to an individual person—nothing. The anticipated trajectory of a tree's life can change at any time for any number of reasons. Its health depends on the stability of the forest ecosystem. It's better if temperature, moisture, and light conditions don't change abruptly, because trees react extremely slowly. But even when all the external conditions are optimal, insects, fungi, bacteria, and viruses are always lurking, waiting for the chance to strike. That usually happens only when a tree gets out of balance. Under normal circumstances, a tree carefully apportions its energy. The largest portion is used for daily living: the tree has to breathe, "digest" its food, supply its fungal allies with sugar, and grow a bit every day. Then the tree has to keep hidden reserves of energy on hand to fight off pests.

These secret reserves can be activated at any time, and depending on the tree species, they contain a selection of defensive compounds produced by the tree. These so-called phytoncides have antibiotic properties, and there has been some impressive research done on them. A biologist from Leningrad, Boris Tokin, described them like this back in 1956: if you add a pinch of crushed spruce or pine needles to a drop of water that contains protozoa, in less than a second, the protozoa are dead. In the same paper, Tokin writes that the air in young pine forests is almost germfree, thanks to the phytoncides released by the needles.[56] In essence, then, trees disinfect their surroundings. But that isn't all. Walnuts have compounds in their leaves that deal so effectively with insects that garden lovers are often advised to put a bench under a canopy of walnuts if they want a comfortable place to relax in the garden, because this is where they will have the least chance of being bitten by mosquitoes. The phytoncides in conifers are particularly pungent, and they are the origin of that heady forest scent that is especially intense on hot summer days.

If the carefully calibrated balance of energy for growth and defense gets thrown out of alignment, then a tree might get sick. This can happen, for example, when a neighboring tree dies. Suddenly, the crown gets more light, and now what the tree wants more than anything is more photosynthesis. That makes sense because a chance like this comes along only about once every hundred years. The tree, finding itself suddenly bathed in sunlight, forgets about everything else and focuses exclusively on growing branches. It has no option really, because its surrounding cohort is doing the same thing, which means that the gap in the canopy will close again in about twenty years, which, if you are a tree, means you don't have much time.

Suddenly, growth speeds up, and instead of adding a few fractions of an inch each year, the tree is adding about 20 inches. This takes energy, which is then not available for fending off illnesses and pests. If the tree is lucky, all goes well, and once the canopy closes again, the tree will have increased the size of its crown. Then it will take a break and settle back into apportioning its energy in a way that suits its lifestyle. But woe betide the tree if something goes wrong during this growth spurt. A fungus might attack the stub left by a fallen branch and, unnoticed, make its way along the dead wood and into the trunk. Or a bark beetle might take an exploratory bite out of a tree busy reaching for the light and discover there is no defensive response. Then the game is up. The trunk, which appears to be in the very best of health, finds itself increasingly under attack because it doesn't have the energy to mobilize a defense.

The first reactions to the attacks soon show up in the treetops. In deciduous trees, the vital topmost growth suddenly dies, leaving thick branch stubs with no side branches sticking up into the sky. The initial reaction of conifers is that their needles don't last as long. Sick pines, for instance, retain not three but maybe only one or two generations of needles on their branches, which makes their crowns noticeably more open. In spruce you also get what is known in Germany as the "Lametta effect," where the twigs hang limply from the branches. (Lametta is another name for the tinsel that is draped over the branches of Christmas trees.) A short time later, big flakes of bark break off the trunk. Things can deteriorate quite quickly from this point. Like a deflating hot air balloon, the crown implodes and sinks as it dies, and winter storms break off the dead branches. You can see this even more clearly with spruce, because the desiccated tips of dying trees contrast clearly with the living green of the lower branches.

Every year, a live tree adds a growth ring to the wood in its trunk because it is, you could say, damned to grow whether it wants to or not. In the growing season, the cambium, that narrow layer of clear cells between the bark and the wood, grows new woody cells on the inside and new bark cells on the outside. If a tree cannot increase its girth, it dies. At least, that is what we thought for a long time. Then researchers noticed pines in Switzerland that looked outwardly healthy and were covered in green needles. On closer inspection—either by cutting the trees down or taking core samples—researchers discovered that a few of them hadn't created a single new growth ring for more than thirty years.[57] Dead pines covered in green needles? The trees had been attacked by an aggressive fungus called annosus root rot, and their cambium had died. But the roots were still pumping water up to the crown through the long narrow transport vessels in the trunk, providing the needles with life-giving moisture. And the roots themselves? When the cambium is dead, the bark is too, which means the tree can no longer pump sugar solutions from its needles back down to its roots. Therefore, healthy neighboring pines must have been helping their dying comrades by supplying their roots with food, as I've described in chapter 1, "Friendships."

Apart from diseases, a lot of trees suffer injuries over the course of their lives. There are many different ways this can happen. A neighboring tree might fall. In a dense forest, a falling tree cannot avoid hitting a few surrounding comrades. If this happens in winter, when the trees' relatively dry bark fits tightly around their wood, not much happens. Most often only a few branches break, leaving no visible signs of damage after just a few years. Injuries to the trunk are more serious, and these usually happen during the summer months. This is when the cambium is full of water, crystal clear, and slippery. At this time, it doesn't take much pressure to loosen the outer layer of the tree, and branches from a falling neighbor scraping by can rip yard-long wounds in the tree's trunk. Ouch! The damp wound is an ideal landing site for fungal spores, which arrive just minutes later. They grow fungal threads that immediately begin making a meal of the wood and the tree's food supplies. But they don't make much progress. There's simply too much water in the wood. Although fungi like it moist, soaking wet conditions spell death for them. At first, their victory march into the interior of the trunk is slowed by the wet outer wood where sap is flowing; however, the sapwood is now exposed to the air, and its outer layers can dry out. A slow-motion struggle begins.

The fungus advances as the sapwood loses its moisture, while the tree tries to close up the wound. To do this, the tissues around the injury really get going and start growing together as fast as they can. They can cover up to a third of an inch of injured wood per year. To avoid complications in the future, the tree must get the area sealed once again within five years, tops. Once new bark has closed the old wound, the tree can saturate the damaged sapwood from the inside and kill the fungus. However, if the fungus has made it from the sapwood into the heartwood—the older wood beneath the sapwood that no longer transports water or stores energy reserves—then it's too late. The decommissioned wood is drier and, therefore, ideal for the attacker, and the tree can't mount any defenses here. So, the decisive factor in whether the tree has a chance is the size of the wound. Anything much more than an inch is life threatening.

But even if the fungus wins and makes itself at home inside the tree, all is not yet lost. True, the fungus can get stuck into the wood without further hindrance, but it takes its time. A whole century can pass before everything has been consumed and turned to mush. Even this won't make the tree the slightest bit less stable, because the fungus cannot expand into the wet outer growth rings of the living sapwood. In extreme cases, the tree gets hollowed out like a stovepipe. And just like a pipe, the tree remains stable. So we shouldn't feel sorry for a rotten tree, and it doesn't necessarily feel pain either, because the heartwood is no longer active and usually no longer contains any living cells. The outer growth rings, which are still active, transport water up the trunk and, therefore, are much too wet for fungi.

If a tree has successfully walled off—that is to say, closed up—an injury to its trunk, then it can usually grow as old as its uninjured companions. But sometimes, especially in cold winters, the old wounds can act up again. Then a crack like a rifle shot echoes through the forest and the trunk splits open along the old injury. This is caused by differences in tension in the frozen wood, because the wood in trees with a history of injury varies greatly in density.

26

— LET THERE BE LIGHT —

I'VE ALREADY TALKED a lot about sunlight, and it's turned out to be an extremely important factor in the forest. This should come as no surprise. After all, trees are plants and need to photosynthesize to survive. But because enough sun usually shines on our garden beds and lawns, in the home garden, water and fertile soil tend to be more decisive factors for plant growth. In our everyday lives, we don't notice that light is more important, and because we like to apply our own situations to others, we overlook the fact that an intact forest has completely different priorities.

In the forest, there's a battle for every last ray of sunlight, and each species is specialized to grow in a particular niche so that it can soak up some energy, however paltry the amount might be. In the upper story—the executive offices—the mighty beeches, firs, and spruce stretch out and soak up 97 percent of the sunlight. This behavior is cruel and inconsiderate, but doesn't every species take what it can? Trees have won this competition for the sun because they grow such tall trunks. But a plant can grow a long sturdy trunk only if it lives for a very long time, because an enormous amount of energy is stored in its wood. To grow its trunk, a mature beech needs as much sugar and cellulose as there is in a 2.5-acre field of wheat. Of course, it takes not 1 but 150 years to grow such a mighty structure, but once it's up there, hardly any other plants—except for other trees—can reach it, and the rest of its life is worry free. Its own offspring are designed to survive in what light remains, and of course, their mothers feed them as well. That is not the case for the rest of the rank and file, and they must come up with other strategies for survival.

Some plants bloom early. In April, seas of frothy white blooms cover the brown earth under old deciduous trees as wood anemones cast their spell on the forest. Sometimes yellow or violet-blue flowers are mixed in, such as liverleafs, so-called because their leaves are shaped a bit like human livers. They earned one of their common names in German—*Vorwitzchen*, or "cheeky little ones"—because the flowers appear so early in the year. Liverleafs are stubborn plants. Once they've found a spot, they want to stay there forever, and they spread very slowly by seed. That's why you find these early bloomers only in deciduous woods that have been around for at least a few hundred years.

The colorful troupe just about exhausts itself putting on a glorious floral show. The reason for this extravagance is that they want to make the most of the short window of time available to them. While the spring sun warms the forest floor from March to early May, the deciduous trees sleep on. Under the giants' bare branches, Liverleafs & Co. seize the opportunity to produce the carbohydrates they need for the following year. They store the food in their roots. In addition, the little beauties have to reproduce, which uses up additional energy. It's a small miracle they can pull it all off in just a month or two. As soon as the buds break on the trees, it gets much too dark again, and the flowers are forced to take another ten months off.

When I said earlier that hardly any other plant can reach the tree's height, the emphasis was on "hardly." For there are some plants that can make it up into the canopy. It's particularly arduous and tedious to start right from the bottom. Ivy is one plant that does this. Ivy begins as a small seed at the foot of a tree with an open growth habit—those species that are particularly wasteful with sunbeams and allow any number of them to fall to the forest floor unused. Under pines or oaks, that's enough for a nice thick carpet of ivy to grow—at first, just on the forest floor. Then, one day, a tendril starts to climb up a trunk. Ivy is the only plant in Central Europe that uses small aboveground roots to anchor itself firmly to bark. Over the course of many decades, the ivy keeps climbing upward until it finally reaches the crown. It can live many hundreds of years up here, though ivy that old is more often found on rocky cliffs or castle walls. Some of the European literature suggests that ivy doesn't hurt the trees it grows on. After observing the trees growing around our house, I can't support this view. Quite the opposite, in fact. Pines need a lot of light for their needles, and they particularly resent this competitor taking over in the treetops. Branches begin to die, and this can weaken trees so much that they give up. Ivy vines encircling trunks can grow as thick as small trees, and like boa constrictors winding themselves around their victims, they can squeeze the life out of pines and oaks.

The process of strangulation is even more apparent in another species: the honeysuckle. This plant, with its pretty lilylike flowers, prefers to climb up younger trees. The honeysuckle wraps itself so tightly around the little trunks that as they grow, they develop deep spiral-shaped indentations. As I've mentioned already, people like to sell these deformed trees as bizarrely shaped walking sticks, which is fine as the trees wouldn't have survived much longer out in Nature anyway. Because their growth has been slowed, trees hugged by honeysuckle fall behind the other youngsters. Even if they do manage to grow up, sooner or later a passing storm will break their twisted trunks.

Mistletoes save themselves the arduous task of climbing up trees. They prefer to start at the top. To do this, they co-opt thrushes, who deposit the mistletoes' sticky seeds when they clean off their beaks on the upper branches. But how do plants survive up there with no contact with the ground to get water or food? Now, way up in those lofty heights, there's water and food aplenty—in the trees. To get at them, the mistletoes sink their roots into the branches they're sitting on and simply suck out what they need. They are photosynthesizing for themselves, at least, so the host tree is "only" short water and minerals. That's why scientists call them "hemiparasites" and not true parasites. But that's not much help to the tree. Over the years, the number of mistletoes in its crown multiplies. You can recognize affected trees—deciduous trees, anyway—in the cold season. Some trees are absolutely covered with these parasitic plants, and in large quantities they can be dangerous. The constant bloodletting weakens the tree, which, incidentally, is also getting increasingly robbed of light. And as if that were not enough, the mistletoe roots massively weaken the structure of the wood in the branches, which often break after a few years, reducing the size of the crown. Sometimes it all gets too much, and the tree dies.

Other plants that simply use the trees for support are less damaging. These would be the mosses. Many species have no roots to sink into soil, or branches; instead they have small hairlike structures, and these are what they use to hold on to the bark. Very little light, no nutrient uptake, no water from the ground, and no tapping of the tree for help: how does that work? It only works if you are extremely frugal. The soft cushions of moss catch water from mist, fog, or rain and store it. Often that is not enough, as the trees either act like umbrellas (Spruce & Co.) or their branches funnel the water down to their roots (deciduous trees). In the latter case, the solution is simple: mosses move into places on the trunk where the water trickles down after a shower. It's not an even distribution because most trees are tilted slightly to one side. A small stream forms on the upper side of a slight bend, and that's what the moss taps into. Incidentally, that is why you can't rely on moss if you want to figure out compass directions. In climates where there is rain year round, moss supposedly indicates the weather side of the tree, where the trunk gets wet when the rain hits it; however, in the middle of the forest, where the wind is stilled, rain usually falls vertically. In addition, each tree is bent in a slightly different direction, so if you were to orient yourself according to moss, you'd only end up confused.

If the bark is rough as well, moisture remains in its tiny fissures for a particularly long time. Rough bark begins at the bottom of the tree and keeps moving upward in the direction of the crown as the tree ages. That's why you find moss growing only an inch or two above the ground on young trees, whereas later it encases the lower trunk like a knee-high sock. Moss doesn't damage the tree, and the tiny plants compensate for the small amount of water they divert by releasing moisture as well, so their influence on the forest climate is positive.

We're left with the question of where moss gets its food. If food doesn't come from the ground, the only place it can come from is the air. And a whole lot of dust is blown through forests every year. A mature tree can filter out more than 200 pounds, which rain flushes down the trunk. Mosses soak up the dusty mixture and filter out what they can use. That deals with the food, and now the only thing missing is light.

In bright pine or oak forests, light isn't a problem, but it is in those eternally dark spruce forests. Even the most abstemious must give these a miss, and that's why particularly dense stands of young trees in coniferous forests are most often completely moss free. It is only as the trees age, when here and there gaps appear in the canopy, that enough sunlight filters through for the trees to get a covering of green. Things are rather different in old beech forests, for here the mosses benefit from leaf-free interludes in spring and fall. It gets too dark again in summer, but the plants are adapted to cycles of hunger and thirst. Sometimes there's no rain for months on end. If you run your fingers over a cushion of moss in a dry spell, you'll find it is completely desiccated. Most plants would die at this stage, but not moss. It swells with the next heavy rain shower—and life continues.

Lichen are even more frugal. These small gray-green growths are a symbiotic combination of fungi and algae. To hold together, they need some kind of a substrate, and in the forest, this is provided by trees. In contrast to moss, they climb much higher up the trunks, where their already extremely slow growth is slowed still further by the leafy canopy. Often it takes them many years to grow a moldy-looking coating over the bark, which prompts many visitors to my forest to ask whether the trees are sick. The trees are not sick; lichen doesn't do them any harm, and the trees are probably completely indifferent to their presence. These tiny growths balance their snail's pace when growing with extreme longevity. They can survive to be hundreds of years old, showing that these organisms are perfectly suited to the slow rhythms of life in ancient forests.

27

— STREET KIDS —

HAVE YOU EVER wondered why giant redwoods in Europe never grow particularly tall? Even though quite a lot of them are more than 150 years old, very few have yet topped 160 feet. In their homeland—forests on the western slopes of the Sierra Nevada mountains in California—they easily grow more than twice that size. Why don't they do that in Europe? If we think back to tree kindergarten, to their extremely long and drawn-out youth, we might be tempted to say: They're still children. What do you expect? But that doesn't jibe with the enormous diameters of the older giant redwoods in Europe, which often exceed 8 feet (measured at chest height). Clearly, they know how to grow. They just seem to be putting their energy into growing in the wrong direction.

Their location gives a clue as to why this might be the case. They were often planted in city parks by princes and politicians as exotic trophies. What is missing here, above all, is the forest, or—more specifically—relatives. At 150 years old, they are, when you consider a potential life-span of many thousands of years, indeed only children, growing up here in Europe far from their home and without their parents. No uncles, no aunts, no cheerful nursery school—no, they have lived all their lives out on a lonely limb. And what about the many other trees in the park? Don't they form something like a forest, and couldn't they act like surrogate parents? They usually would have been planted at the same time and so could offer the little redwoods no assistance or protection. In addition, they are very, very different kinds of trees. To let lindens, oaks, or beeches bring up a redwood would be like leaving human children in the care of mice, kangaroos, or humpback whales. It just doesn't work, and the little Americans have had to fend for themselves. No mother to nurse them or keep a strict eye out to make sure the little ones didn't grow too quickly. No cozy, calm, moist forest around them. Nothing but solitude.

And if that weren't enough, in most cases, the soil is a complete disaster. Whereas the old-growth forest offers soft, crumbly, humus-rich, and constantly moist soil for their delicate roots, European parks offer hard surfaces that have been depleted of nutrients and compacted after years of urbanization. What's more, members of the public like to walk up to the trees, touch their bark, and relax in the shadow of their crowns. Over the decades, constant trampling around the base of the trees leads to further soil compaction, which means that rain drains away far too quickly, and in winter, the trees cannot build up a supply of water to last the summer.

The mechanics of planting also haunt the trees for the rest of their lives. They are kept alive and handled in nurseries for years before being moved to their final locations. Every fall, their roots are trimmed to keep them compact in the nursery beds so that they can later be moved more easily. The root ball, which for a 10-foot-tall tree grows to about 20 feet in diameter if left to its own devices, is cut back to about 20 inches, and to make sure the crown doesn't wilt from thirst thanks to the root reduction, it too is heavily reduced. All this is done not to improve the health of the tree but simply to make it easier to handle. Unfortunately, when the roots are pruned, the brain-like structures are cut off along with the sensitive tips. Ouch! After that, it is as if this interference makes the trees lose their sense of direction underground. They stop growing roots down into the soil and form a flat plate of roots near the surface instead, severely restricting the trees' ability to find water and food.

At first, the young trees don't seem to mind. They stuff themselves with sugary treats because they can photosynthesize as much as they like in full sun. It's so easy to get over the loss of a mother's tender care. And in the early years, the water problems in a rock-hard soil are barely noticeable. After all, the saplings are being lovingly cared for and watered by gardeners when they get dry. But above all, there is no strict discipline. No "Take it easy," no "Just wait a couple of hundred years," no punishing light deprivation if you don't grow up really straight. Every young tree can do just as it likes. So, every year, they go at it as though they were in a race, and every year, they put on a growth spurt. After a certain height, the childhood bonus seems to run out. Irrigating 65-foot-tall trees takes an enormous amount of water and time. To thoroughly moisten the roots, the gardeners must spray many gallons of water out of their hoses—per tree! And so, one day, the care simply stops.

At first, the giant redwoods don't really notice. They've lived high on the hog for decades and done whatever they wanted. Their thick trunks are like paunches attesting to an orgy of solar indulgence. In the early years, it doesn't really matter much that the cells inside their trunks are very large, contain a lot of air, and therefore are susceptible to fungal infections. Their side branches also show signs of their loutish behavior. The trees in the park know nothing about the etiquette manual that guides the old-growth forest, calling for thin branches in the lower regions of the trunk, or even for no branches at all. Thanks to the generous amounts of light that reach right to the ground, the redwoods grow thick side branches that later increase their girth so much that the image the trees bring to mind is that of doped-up body builders. True, all the branches on the lower 6 to 10 feet of the trees are usually sawn off by the gardeners to give visitors an unobstructed view of the park, but when compared with old-growth forests, where thicker branches are not allowed below 65 and sometimes not even below 165 feet, the trees' growth is brazenly decadent.

What the trees end up with are short, thick trunks topped with crowns. Extreme examples of park trees seem to be nothing but crown. Their roots don't penetrate more than 20 inches down into the heavily trampled soil, and therefore, they offer little in the way of support. That's very risky, and trees of a normal height would be much too wobbly. The growth habit of redwoods in the far-off old-growth forests ensures a low center of gravity, so they are pretty stable. It takes a huge storm to upset their equilibrium.

Once European redwoods have passed the hundred-year mark (the trees are now the age of schoolchildren), that's the end of easy living. The topmost branches wither away, and no matter how hard the trees try to grow up again, they have reached the end of the road. Their wood is impregnated with natural fungicides, so they can hold out for many more decades despite injuries to their bark. It's quite different with other species of tree. Beeches, for example, react badly when thick branches are sawn off. Take a closer look the next time you take a walk in a park. You'll find hardly any large deciduous trees that don't show signs of having branches trimmed, sawn off, or interfered with in some other way. This "pruning" (it's actually more like a massacre) is often only for aesthetics, which dictates that the crowns of trees lining a walk or driveway are all the same size and shape.

A severely pruned crown is a severe blow for the roots, which grow to a size optimally suited to serve the aboveground parts of the tree. If a large percentage of the branches is removed and the level of photosynthesis drops, then just as large a percentage of the underground part of the tree starves. Fungi now penetrate the dead ends where branches have been removed and the trunk has been sawn off. The wood is filled with air pockets, thanks to the tree's quick growth as a youngster, and fungi have a field day. After only a few decades—which is incredibly fast for a tree—this inner rot can also be seen on the outside of the tree. Complete sections of the crown die off, until the local authorities cut the crown off completely so that it no longer poses a safety hazard for visitors, leaving huge wounds where the tree has been topped. The waxy substance painted over the damaged trunk, supposedly to protect it, often hastens the tree's demise because it traps moisture inside, creating the damp conditions fungi love. In the end, all that remains is an empty shell that cannot be saved and one day will be chopped down. And because there are no family members who can rush to help these urban trees, the stump will die quickly and completely. A little while later, a new tree will be planted and the drama will begin all over again.

Urban trees are the street kids of the forest. And some are growing in locations that make the name an even better fit—right on the street. The first few decades of their lives are similar to those of their colleagues in the park. They are pampered and primped. Sometimes they even have their own personal irrigation lines and customized watering schedules. When their roots want to go out and get established in their new territory, they're in for a big surprise. The soil under the street or pedestrian walkway is harder even than the soil in parks, because it has been compacted by machines using large vibrating metal plates. That's a huge disappointment for the tree. The roots of forest trees don't actually grow very deep. Few species grow deeper than 5 feet, and most call a halt to downward growth much sooner. That's not a problem in the forest, where there is almost no limit as to how wide the roots can grow. Unfortunately, this isn't the case on the side of the street. The roadway restricts growth in one direction, there are pipes under the pedestrian zones, and soil has been compacted during construction.

When trees are planted in these restricted spaces, conflicts are inevitable. In such places, plane trees, maples, and lindens like to feel out underground wastewater pipes. We notice the damage when the next storm comes and the streets fill with water. Then specialists armed with root probes investigate to see which tree has caused the blockage. The culprit is sentenced to death for its excursion under the sidewalk and into what it thought was paradise. The offending tree is cut down, and its successor is planted in a built-in root cage to discourage such behavior in the future.

Why do trees grow into pipes in the first place? For a long time, city engineers thought the roots were somehow attracted by moisture seeping from loose connections between the pipes or by nutrients in the wastewater. However, the results of an extensive applied study by the Ruhr University Bochum point in a completely different direction. The study found the roots in the pipes were growing above the water table and did not seem interested in extra nutrients. What was attracting them was loose soil that had not been fully compacted after construction. Here, the roots found room to breathe and grow. It was only incidentally that they penetrated the seals between individual sections of pipe and eventually ran riot inside them.[58] What this means is that when trees in urban areas run up against ground as hard as concrete wherever they turn, they get desperate, and it is only as an absolutely last resort that they finally find a way out into sloppily backfilled trenches. Once they get there, they are a problem.

There is no remedial support for the trees, only for the pipes, which are now reburied in especially well-tamped-down soil so that the tree roots can no longer find a footing there. Are you surprised that summer storms topple a particularly large number of street trees? Their puny underground anchoring systems—which in Nature could cover more than 700 square yards and are now restricted to an area shrunk to a tiny percentage of that—are not capable of supporting trunks that weigh many tons.

But there is even more these tough plants have to bear. The urban microclimate is heavily influenced by heat-inducing asphalt and concrete. Whereas forests cool themselves on hot summer nights, streets and buildings radiate the heat they soaked up during the day, keeping temperatures elevated. Radiated heat makes the air extremely dry. Not only that, but it's full of exhaust fumes. Many of the companions that look after trees' well-being in the forest (such as the microorganisms that make humus) are missing. Mycorrhizal fungi that help collect water and food are present only in low numbers. Urban trees, therefore, have to go it alone under the harshest conditions.

As if that were not enough, they also have to deal with unsolicited extra fertilizers. Above all, from dogs, which lift their legs at every available trunk. Their urine can burn bark and kill roots. Winter salt leads to similar damage. Depending on the severity of the cold, salt is sometimes applied around trees at the rate of 2.2 pounds per square yard. In addition, the needles on conifers, which are still attached to the branches in winter, have to deal with the salt spray thrown up by car tires. At least 10 percent of the salt ends up in the air and falls back down on trees—among other resting places—where it burns the foliage. These painful injuries show up as small yellow and brown spots on the needles. The burns reduce the trees' ability to photosynthesize the next summer and, therefore, weaken the trees.

Weakness equals pests. It's easier for scales and aphids to strike, because street trees have limited resources they can put toward defending themselves. High urban temperatures are a contributing factor. Hot summers and warm winters favor the insects, which survive in larger numbers. In Central Europe, one species constantly makes headlines because it is a menace to the human population as well: the oak processionary. This moth gets its name because after feeding in the crown, its caterpillars crawl nose to tail down the trunk in long lines. They protect themselves from predators using thick webs, where they retreat to molt as they grow. People fear the little pests because they are covered in fine stinging hairs, which break off when you touch them and make their way under the skin. There, like stinging nettles, they release substances that itch and cause welts and can even trigger acute allergic reactions. The stinging hairs on the shed skins remain hanging in the webs and can inflict damage for up to ten years. In urban areas, the arrival of these insects can spoil a whole summer, yet ultimately, they are not the ones at fault.

The oak processionary is relatively rare in Nature. Just a few decades ago, it was on the list for critically endangered species, and now everyone everywhere wants to get rid of them. Population explosions have been described for more than two hundred years. The German Federal Agency for Nature Conservation doesn't attribute these infestations to climate change and rising temperatures but to the presence of attractive food sources for the moth.[59] They love warm crowns drenched in sunlight. In the middle of the forest, these are hard to find. The few oaks that grow in the forest are mixed in with beeches, and only their topmost tips reach the light. In the city, however, oaks stand out in the open, where they are warmed by the sun all day long. The caterpillars love this. And as the whole "forest" in urban areas offers such perfect conditions, it's no surprise that there are population explosions, which are a stern reminder that oaks and other species growing along the streets and between houses have to fight for their lives.

At the end of the day, the stresses the trees must bear are so great that most of them die prematurely. Even though they can do whatever they want when they're young, this freedom is not enough to compensate for the disadvantages they face later in life. One consolation is that because streets and pathways are often planted with rows of the same species of trees, at least they are able to communicate with other members of their species. Plane trees—recognizable by their attractive bark, which peels off in colorful flakes—are a popular choice for these regimented plantings. Whatever it is these street kids talk to each other about through their scent-mail—and whether the tone of these messages is as rough as their lives—the street gangs are keeping this information strictly to themselves.

28

— BURNOUT —

Street kids are denied the cozy atmosphere of the forest. And because they are trapped where they have been planted, they have no choice. There are, however, a few species of tree that couldn't care less about the forest's comforts and social interactions and prefer to strike out on their own. These are the so-called pioneer tree species (that sounds much better), which like to grow up as far away from their mothers as they can. Accordingly, their seeds are capable of flying long distances. They are very small and padded or equipped with tiny wings so that powerful storms can carry them for miles. Their goal is to land outside the forest, where they can colonize new areas.

The site of a devastating landslide or a recent volcanic eruption that spewed enormous quantities of ash, areas torched by forest fire—all have potential as long as there aren't any large trees there already. There's a reason for this: pioneer species hate shade. Shade slows their upward growth, and a tree that grows slowly has already lost. A race for a place in the sun erupts among the Johnny-on-the-spots. These eager beavers include different species of poplar, such as quaking aspen, and silver birch and pussy willow. In contrast to small beeches and pines, whose annual growth is measured in fractions of an inch per year, the pioneers sometimes grow more than 3 feet taller in the same period. In just ten years, they can transform land that once lay fallow into a young forest rustling in the breeze. And most of these quick starters are blooming by then to give their seeds a head start in the search for new realms to conquer and a chance to occupy the last remaining open patches of ground around them.

An open space, however, is attractive to herbivores, because it's not only trees that try their luck here but also grasses and wild flowers, which don't do well in the forest understory. Deer—or, in earlier times, wild horses, aurochs, and bison—are drawn to these plants. Grasses are adapted to constant grazing and are relieved that the young trees that threaten their existence are being polished off in the process. Many shrubs that would dearly love to grow taller than the grasses have developed dangerous thorns to protect themselves from the voracious beasts. Blackthorn is so vicious that its pointed protrusions persist on dead plants for years to impale rubber boots and even car tires, to say nothing of the hides and hooves of animals.

Pioneer trees seek to defend themselves in other ways, as well. They grow quickly, so their trunks get thick fast, and they put on a massive layer of rough outer bark. You can see evidence of this rapid growth on silver birches, where black fissures split their smooth white exteriors. Not only do browsers break their teeth on the tough bark, but they are also revolted by the taste of its oil-saturated fibers. This oil, by the way, is the reason even green birch bark burns so wonderfully well and is great for lighting campfires. (If you're going to try this, pull off only the outermost layer of bark so that you don't harm the tree.)

Silver birch bark has another surprise in store. The white color is because of the active ingredient betulin, its primary component. White reflects sunlight and protects the trunk from sunscald. It also guards the trunk against heating up in the warming rays of the winter sun, which could cause unprotected trees to burst. As birches are pioneer trees that often grow all alone in wide-open spaces without any neighbors to shade them, such a feature makes sense. Betulin also has antiviral and antibacterial properties and is an ingredient in medicines and in many skin care products.[60]

What's really surprising is how much betulin there is in birch bark. A tree that makes its bark primarily out of defensive compounds is a tree that is constantly on the alert. In such a tree there is no carefully calibrated balance between growth and healing compounds. Instead, defensive armoring is being thrown up at a breakneck pace everywhere. Why doesn't every species of tree do that? Wouldn't it make sense to be so thoroughly prepared against attack that potential aggressors would breathe their last the moment they took the first bite? Species that live in social groups don't entertain this option because every individual belongs to a community that will look after it in times of need, warn it of impending dangers, and feed it when it is sick or in distress. Cutting back on defense saves energy, which the tree can then invest in producing wood, leaves, and fruit. Not so with the birches, which must be completely self-reliant if they are to survive. But they, too, grow wood—and indeed, they do so a lot faster—and they, too, want to, and do, reproduce. Where does all their energy come from? Can this species somehow photosynthesize more efficiently than others? No. The secret, it turns out, lies in wildly overtaxing their resources. Birches rush through life, live beyond their means, and eventually wear themselves out. But before we take a look at the results of this behavior, allow me to introduce you to another unsettled spirit: the quaking aspen.

The quaking aspen takes its name from its leaves, which react to the slightest breath of wind. And although we have sayings that associate this characteristic with fear ("to shake like a leaf"), quaking aspens don't shake because they are afraid. Their leaves hang from flexible stems and flutter in the breeze, exposing first their upper and then their lower surfaces to the sun. This means both sides of the leaf can photosynthesize. This is in contrast to other species, where the underside is reserved for breathing. Thus, quaking aspens can generate more energy, and they can grow even faster than birches.

When it comes to predators, the quaking aspen pursues a completely different strategy from the birch, relying on stubbornness and size. Even when they are being nibbled down by deer year after year, they slowly expand their root systems. From their roots, they then grow hundreds of subsidiary shoots, which, as the years progress, develop into decent-sized trunks. Accordingly, a single tree can extend over many hundreds of square yards of ground—or, in extreme cases, even farther. In Fishlake National Forest, Utah, there is a quaking aspen that has taken thousands of years to cover more than 100 acres and grow more than forty thousand trunks. This organism, which looks like a large forest, has been given the name "Pando" (from the Latin "*pandere*," which means to spread).[61] You can see something similar in forests and fields in Europe, albeit not on such a grand scale. Once the brush has become sufficiently impenetrable, then a few of the trunks can grow upward undisturbed and develop into large trees in less than twenty years.

It goes without saying that constant struggle and rapid growth exact their toll. After the first three decades, exhaustion sets in. The topmost branches, a yardstick for the vitality of pioneer tree species, thin out. That in itself wouldn't be too worrisome, but trouble is brewing under the poplars, birches, and willows. Because they let a lot of light shine through their crowns and reach the ground unused, Johnny-come-latelies can get a foothold. These would be the slower-growing maples, beeches, hornbeams, or even silver firs, which prefer to spend their childhoods in the shade anyway. The pioneer species have no choice but to shade them, and when they do, they are signing their own death warrants. A competition begins that they will, inevitably, lose. The interloping youngsters gradually grow taller, and after a few decades, they catch up with the trees affording them shade. By this time, their benefactors are burned out, completely spent, and top out their growth at a maximum of 80 feet.

For Beeches & Co., 80 feet is nothing. They weave their way through the crowns of the pioneer trees and happily grow up and out over them. With their dense crowns, they are considerably better at exploiting the light, and now not enough of this precious commodity reaches the birches and poplars they have overtaken. The distressed trees put up a fight, especially the silver birches, which have developed a strategy to keep the troublesome competition at bay for at least a few more years: their long, thin, pendulous branches act like whips, and they lash out in all directions in even the lightest breeze. This whipping action damages the crowns of neighboring non-related trees, slaps off their leaves and new growth, and, at least in the short term, restricts their growth. Despite this, the lowly tenants eventually overtake the birches and poplars and now everything happens relatively quickly. After just a few years, their last reserves used up, the pioneer species die and return to humus.

But their lives would be relatively short compared with other forest trees even without the hard-hitting competition. As their upward growth slows, their defenses against fungi disappear. One broken-off branch is enough to provide a port of entry. Because their wood is composed of large cells grown in haste, it contains a lot of air, and so the destructive fungal filaments can spread quickly. The trunk rots big time, and because pioneer species often stand out in the open alone, it's not long until the next fall storm topples the tree. This is not a tragedy for the species itself. Its goal of rapid dispersal was achieved a long time ago, as soon as it quickly reached sexual maturity and propagated.

29

— DESTINATION NORTH! —

TREES CAN'T WALK. Everyone knows that. Be that as it may, they need to hit the road somehow. But how can they do this without feet? The answer lies in the transition to the next generation. Every tree has to stay where it put down roots as a seedling. However, it can reproduce, and in that brief moment when tree embryos are still packed into seeds, they are free. The moment they fall from the tree, the journey can begin.

Some species are in a big hurry. They equip their offspring with fine hairs so that they can drift off on the next wind, light as a feather. Species that rely on this strategy have to grow tiny seeds so that they are light enough to float away. Poplars and willows produce minute fliers like this and send them off on half-mile-long journeys. The advantage of long-distance travel is offset by the disadvantage that the seeds contain hardly any provisions. The sprouting seed quickly uses up its energy reserves, making it highly susceptible to starvation and thirst. The seeds of birches, maples, hornbeams, ash, and conifers are somewhat heavier. At this weight, flight in a feathery coating is no longer practical, so these trees equip their fruit with flying aids. Some species, such as conifers, have an efficient winged design for their seeds, which works well to slow the seeds as they fall. If a storm blows through when the seeds are falling, they can travel about a mile. Species that produce heavy fruit, such as oaks, chestnuts, or beeches, could never cover such distances. Therefore, they avoid any kind of structural assistance and instead enter into an alliance with the animal world.

Mice, squirrels, and jays love oily, starchy seeds. They tuck them into the forest floor as winter provisions, and there the seeds often stay, lost or no longer needed. Sometimes a hungry tawny owl swoops down and a yellow-necked mouse ends up as a meal itself. And so the little rodent makes its contribution to the next generation of trees, small though it might be. These mice often bury their winter stores directly at the base of the trunk of the mighty beech whose nuts they gather. There are lots of small dry holes among the roots, and little creatures love to live in them. If a mouse has moved in, you'll find husks of completely consumed beechnuts piled in front. At least a few of these stockpiles are buried a few yards from the tree on the open forest floor. After the death of the mouse, they sprout the following spring and become the new forest.

The jay transports heavy seeds the farthest. It carries acorns and beechnuts a few miles away. The squirrel manages only a few hundred yards, whereas mice bury their supplies barely more than 30 feet from the tree. So if you are a heavy-fruited species, you're certainly not going anywhere quickly. However, the large reserves of food in the seed are a cushion to ensure the seedling has a good chance of surviving its first year.

This means that light-seeded poplars and willows can open up new habitats much more quickly—for example, when a volcanic eruption shuffles the cards in the deck of life and the game starts over. But because these trees don't get very old and allow a lot of light to reach the ground, tree species that arrive on the scene later eventually take over. But why make the journey at all? Couldn't the forest just stay right where it is, where things are comfortable and pleasant?

Opening up new places to live is necessary primarily because the climate is always changing. It's changing very slowly, to be sure, over the course of many hundreds of years, but eventually, despite whatever built-in tolerance trees might have, it will become too warm, too cold, too dry, or too wet for a particular species. Then the trees must depart for other climes, and this means packing up and moving. Such a migration is happening in Central European forests right now. The reason is not just climate change, which has already presented us with a 1.4-degree Fahrenheit rise in the average temperature, but also the change from the last ice age to a warmer era.

Ice ages are hugely influential. As the centuries get increasingly colder, trees must retreat to more southerly climes. If the shift takes place slowly over many generations, trees in Central Europe, for example, successfully relocate to the Mediterranean region. But if the ice advances quickly, it buries forests and swallows up species that have been dragging their feet.

In Central Europe 3 million years ago, you could find not only the native beeches we have today but also large-leaved beeches. Although beeches managed to make the leap to southern Europe, the less agile large-leaved beeches died out. One reason for their demise was the Alps. This mountain range forms a natural barrier that blocked the trees' escape route. To cross the Alps, the trees had first to settle high terrain before descending once more to more comfortable elevations. But higher places are too cold for many trees, even in interglacial periods, so the fortunes of many species ended when they reached the tree line. Today, you can no longer find large-leaved beeches in Central Europe, but you can find them in eastern North America, where they are known, simply, as American beeches. (The reference to their large leaves can be found in their Latin name, *Fagus grandifolia*—"grandis" means big and "*folia*" means leaves.) American beeches survived because there is no inconvenient east-west mountain range blocking movement from north to south on the North American continent. They could make their way south without hindrance and then move back north after the ice age was over.

Along with a few other tree species, the beeches of Central Europe somehow managed to make it over the Alps and survive in protected locations until our current interglacial period. The road has been open for these relatively few species for thousands of years, and today they are marching north, still, as it were, following the trail of the melting ice. As soon as the climate warmed up, the germinating seedlings were in luck again. They grew to be mature trees and scattered new seeds that progressed north, mile by mile. The average speed of the beeches' journey, by the way, is about a quarter mile—a year.

Beeches are particularly slow. Their seeds are carried off by jays less often than acorns are, and other species spread themselves using the wind and occupy open areas much more quickly. When the easygoing beeches returned about four thousand years ago, the forest was already occupied by oaks and hazels. That was no big deal for the beeches, and you are already familiar with their strategy. They take a lot more shade than other trees and, therefore, have no difficulty sprouting at their feet. The small amount of light that oaks and hazels allow to reach the ground is sufficient for the tiny conquistadors to keep on growing upward and one day to break through the crowns of the competition. What had to happen, happened. The beeches grew up and over the species that had been there earlier and robbed them of the light they needed to survive. Their merciless triumphal march stretches as far north as southern Sweden today, but it is not over yet. Or, rather, it wouldn't have been over had people not interfered.

When beeches arrived, the European forefathers were beginning to make massive changes to forest ecosystems. They were clearing trees around their settlements to make room for fields for their crops and clear-cutting more areas for livestock. And because even this was not enough, people were simply driving their cattle and pigs into the forest. For beeches, this was catastrophic. Their offspring had to endure centuries at ground level before they were allowed to grow. In those days, their topmost buds were defenseless and at the mercy of browsing animals. Originally, there had been very few mammals around, because dense forests offer little food. Before people arrived on the scene, the odds of beeches hanging out for two hundred years undisturbed and uneaten were high. But then came a constant stream of herders with their hungry livestock gobbling up their tasty buds. In areas where light now fell because trees had been cut down, other species of trees previously overshadowed by the beeches took over. This severely hindered the post-ice age migration of beeches, and to this day, there are areas in Europe they have not yet colonized.

In the past few centuries, hunting has come to European forests as well, which, paradoxically, considerably increased the numbers of deer and wild boar. Thanks to massive feeding programs by hunters, who are mostly interested in increasing the number of antler-bearing stags, the population grew until today it is up to five times its natural level. German-speaking regions have one of the highest concentrations of herbivores in the world, so small beeches are finding it harder than ever to survive. And forestry is restricting their spread, as well. In southern Sweden, where beeches could comfortably grow, it's one spruce or pine plantation after another. Except for a few individual trees, there are hardly any beeches to be found there. But they are ready and waiting. The moment people stop interfering, they will resume their northward migration.

The slowest of the migrants is the European silver fir, the only species of fir native to Germany. Its name comes from its light-gray bark, which makes it easy to distinguish from spruce, which have red-brown bark. The silver fir, like most tree species, waited out the ice age in southern Europe, probably in Italy, the Balkans, and Spain.[62] It migrated from there, following the other trees, at a rate of 300 yards a year. Spruce and pines pulled ahead because their seeds are considerably lighter and better fliers. Even the beeches with their heavy nuts were faster, thanks to the jays.

Apparently, silver firs had developed the wrong strategy because their seeds are not good at flying, even though they are equipped with a small sail to catch the wind, and they are too small to be distributed by birds. Although there are birds that eat the seeds of fir trees, that's of minimal use to the conifers. The nutcracker—which prefers the seeds of the Swiss pine but will eat the seeds of firs—gathers the seeds and stockpiles them. But in contrast with the jay, which hides acorns and beechnuts in soil all over the place, the nutcracker stashes his provisions in protected, dry locations. Even if a bird forgets a seed or two, because there's no water the abandoned seeds never sprout.

Life is hard for silver firs. Whereas most of Central Europe's native trees are well on their way to Scandinavia by now, silver firs have made it only as far as the Harz mountains in northern Germany. But what difference does it make to a tree if it's a few hundred years late? After all, firs tolerate deep shade and can grow under beeches. They gradually insinuate themselves even into established old forests and can eventually grow into mighty trees. Their Achilles' heel is that they are delectable to deer. Right now, these herbivores are preventing silver firs from migrating farther north because in some places they are gobbling up every last seedling.

And why is the beech so competitive in Central Europe? Or to put it another way, if it can prevail so well against all other species in Europe, why isn't it found all over the world? The answer is simple. Its strengths are advantageous only in the region's climatic conditions, which are influenced by the relative proximity of the Atlantic Ocean. Apart from up in the mountains (where beeches don't grow on the upper slopes), temperatures don't fluctuate very much. Cool summers are followed by warm winters, and precipitation is between 20 and 60 inches a year, just the way beeches like it.

Water is one of the key factors for growth in the forest, and this is where the beeches score big time. To produce 1 pound of wood, they need 22 gallons of water. Does this sound like a lot? Most other species of tree need up to 36 gallons, almost twice as much, and that is the deciding factor that enables beeches to shoot up quickly and suppress other species. Spruce are predisposed to guzzle water because in their cool, moist comfort zone in far northern regions, drought is unheard of. In Central Europe, only zones just below the tree line offer the conditions spruce enjoy. It rains a lot here, and thanks to the low temperatures, there's hardly any evaporation. Trees growing at these elevations can afford to waste water. In most lower-lying areas, however, the frugal beeches come out ahead. Even in dry years, beeches put on a decent amount of growth and quickly tower over the heads of the spendthrifts. The offspring of the competition suffocate in the thick layer of leaves on the ground, but the beech seedlings have no problem pushing their way through. Beeches' intensive use of light—which leaves nothing for the other species—and their ability to create for themselves the humid microclimate they enjoy, to build up a good supply of humus on the ground, and to gather water with their branches make them unbeatable in Central Europe today. But only in this part of the world.

As soon as the climate warms up and becomes more Mediterranean, these trees are going to have a hard time. They can't tolerate constantly hot, dry summers and bitterly cold winters, and they will have to step aside for other species, such as oaks. Hot summers and cold winters prevail in Eastern Europe. Although Scandinavian summers are still acceptable, the colder times of the year that far north are also not for the beech. And in the sunny south, they like to settle only the higher elevations where it's not quite so hot. Because of the climate it needs, therefore, the beech is currently trapped in Central Europe. Climate change is making the north warmer, and so, in the future, it will be able to expand its range in this direction. At the same time, it will eventually get so hot to the south that the tree's whole range will shift in a northerly direction.

30

TOUGH CUSTOMERS

SO WHY DO trees live so long? After all, they could grow just like wild flowers: grow like gangbusters for the summer, bloom, set seed, and then return to humus. That would have one definite advantage. Every new generation brings with it the opportunity for genetic modifications. These mutations are most likely to occur during mating and fertilization, and in a world that is constantly changing, adaptation is necessary for survival. For example, mice produce a new generation every few weeks; flies are a lot quicker. Every time hereditary traits are passed down, genes can be damaged, and with a stroke of luck, this damage will introduce a particularly beneficial new characteristic. In short, this is what we call evolution. It helps organisms adapt to changing environmental conditions and, therefore, guarantees the survival of each species. The shorter the interval before the next generation, the more quickly animals and plants can adapt.

Trees seem completely uninterested in this scientifically established imperative. They simply live to be ancient—on average many hundreds, but sometimes even thousands, of years old. Of course, they propagate at least every five years, but this doesn't usually produce a completely new generation of trees. What use is it if a tree produces hundreds of thousands of offspring if they cannot find any vacant posts to fill? As long as their mothers are capturing all the light, nothing much happens at their feet, as I have already explained. Even if the young trees exhibit brilliant new traits, they must often wait centuries before they can bloom themselves and pass these genes along. Quite simply, everything moves along very slowly, and you might expect this to put the trees in an almost impossible situation.

If we look back to recent climate history, it is characterized by abrupt changes. A large construction site near Zurich shows just how abrupt. Workers here came across relatively fresh tree stumps, which, at first, they set aside without paying them any attention. A researcher found them, took samples, and investigated their age. The result: the stumps came from pines that were growing there almost fourteen thousand years ago. Even more amazing, though, were the fluctuations in temperature at that time. In less than thirty years, the temperature dropped as much as 42 degrees Fahrenheit, only to finally rise again by about the same amount. That corresponds to the current worst-case climate change scenario we could potentially face by the end of the twenty-first century. Even the last century in Europe, with the bitterly cold 1940s, the record drought in the 1970s, and the way-too-warm 1990s, was very hard on Nature. Trees employ two strategies to stoically endure these changes: behavior and genetic variability.

Trees exhibit great tolerance for variations in climate. And so the native European beech grows from Sicily to southern Sweden. Apart from the capital *S* at the beginning of the place names, these regions have little in common. Birches, pines, and oaks are also very flexible. But this is not enough to satisfy everything they need to do. When temperatures and rainfall fluctuate, many animals and fungi move from south to north and vice versa. That means that trees must also be able to adapt to unfamiliar pests.

The climate can also change so severely that it falls outside the range the trees can tolerate. And because they have no legs to carry them away and nowhere to turn for help, they have to adapt so that they can deal with the situation themselves. The first opportunity to do this comes at the very earliest stage of life. Shortly after fertilization, when the seeds are ripening in the flower, they react to environmental conditions. If it is particularly warm and dry, appropriate genes are activated. Scientists have proved that under these conditions, spruce seedlings are better able to tolerate warm weather—though they lose the same measure in frost resistance.[63]

Mature trees can adapt as well. If spruce survive a dry period with little water, in the future they are markedly more economical with moisture and they don't suck it all up out of the ground right at the beginning of summer. The leaves and needles are the organs where most water is lost through transpiration. If the tree notices that water is in short supply and thirst is becoming a long-term problem, it puts on a thicker coat. The tree toughens up the protective waxy layer on the upper surface of its leaves. The walls of the cells within the leaves keep them watertight, and the tree increases the thickness of the cell walls by adding extra layers. As the tree battens down the hatches, however, it also has a harder time breathing.

Once a tree has exhausted its behavioral repertoire, genetics come into play. As I've just mentioned, it takes an extremely long time to produce a new generation of trees. This means speedy adaptation is not an option, but other responses are available. In a forest that has been left to its own devices, the genetic makeup of each individual tree belonging to the same species is very different. This is in contrast to people, who are genetically very similar. In evolutionary terms, you could say we are all related. In contrast, the individual beeches growing in a stand near where I live are as far apart genetically as different species of animals. This means each tree has different characteristics. Some deal better with drought than cold. Others have powerful defenses against insects. And yet others are perhaps particularly impervious to wet feet. If climatic conditions change, the first individuals to die will be those that have the hardest time dealing with the new status quo. A few old trees will die, but most of the rest of the forest will remain standing. If conditions become more extreme, one tree species could even be decimated without this being the end of the forest. Usually, a sufficiently large number of trees remain to produce enough fruit and shade for the next generation. I made a calculation for the old beech stands in the forest I manage using available scientific data. Even if we were to have a Spanish-style climate here in Hümmel sometime in the future, an overwhelming number of the trees would cope. The only proviso is that the social structure of the forest is not disturbed by lumber operations so that the forest can continue to regulate its own microclimate for itself.

SPRUCE

31

— TURBULENT TIMES —

IN THE FOREST, things don't always work out according to plan. Even though this ecosystem is immensely stable, often humming along for many centuries with no drastic changes, a natural catastrophe could still throw everything into turmoil. I've already written about winter storms. If a hurricane flattens whole forests, it usually affects commercial spruce or pine plantations. They are often growing on land damaged by machines and so compacted that the roots can't grow down into it to provide good support for the trees. Moreover, in Central Europe, these trees grow much taller than they do in their original home farther north, and they hold on to their needles even in the winter. This means there is a large surface area to catch the wind and a long trunk to intensify the pressure. So the fact that the weak roots don't hold is not so much a catastrophic event as simply a logical one.

But there are storm events in which even natural forests sustain at least localized damage. There are tornadoes whose swirling winds change direction in seconds and overwhelm the trees. These turbulent winds often happen in combination with thunderstorms, which, in Central Europe, occur almost only in summer, so yet another factor comes into play: at that time of year, deciduous trees have leaves on their branches. In the "normal" storm months from October to March, Beech & Co. are naked right down to their branches and, therefore, offer little wind resistance. In June or July, however, trees are not expecting these kinds of problems. If a tornado sweeps through a forest, it slams into the crowns and twists them right off with its raw power. The splintered trunks are left standing as a monument to this atmospheric assault, a lasting testament to the forces of Nature.

Tornadoes are rare events, and therefore, in evolutionary terms, it clearly doesn't make sense to develop a defensive strategy just for them. However, there is another type of damage that happens much more often in connection with thunderstorms: the complete collapse of the crowns of deciduous trees because of heavy rain. When enormous amounts of water land on the leaves in just a few minutes, the trees have to handle loads that weigh many tons. Typically, extra weight from above comes in the form of winter snow, and this falls right through the trees because the leaves are already on the ground by then. In summer, snow is not an issue, and beeches and oaks in full leaf have no problem bearing up under a typical rainfall. Even a downpour is usually fine if a tree has grown normally. Things do, however, get a little dicey for trees if they ignored the etiquette manual while they were growing up and now have structural issues with their trunk or branches.

A typical issue that can lead to branch failure is the so-called hazard beam. The name says it all. A normal branch curves like a bow. It comes out from the trunk, grows upward for a while, and then grows horizontally before gently curving down. This gentle curve does a good job of cushioning the impact of weight from above without breaking. That is extremely important, because the branches of older trees can be more than 30 feet long. This long lever exerts enormous pressure at the point where the branch meets the trunk. Despite the dangers, some trees clearly don't want to follow tried-and-true branch patterns. In these trees, the branches start by pointing away from the trunk, only to then bend and grow upward and continue to hold this course. If a branch that grows in this *J* shape is bent down toward the ground, the force of heavy rain is not absorbed, and the branch breaks because downward pressure compresses the fibers on the underside (that would be the fibers on the outside of the *J* curve) and overextends those on the inside. Sometimes it is the trunk itself that is malformed in this way, and these trees break apart in the torrential rains that often accompany thunderstorms. It's just another tough selection process that eliminates unfit trees from the race.

Other times, the breakdown has nothing to do with structural problems in the tree. The pressure from above is simply too great. Such breakdowns mostly happen in the months of March and April when snow is transformed from feather-light fluff to dead weight. You can estimate the point at which snow becomes dangerous by looking at the clusters of flakes as they fall. When the clusters are about the same diameter as a two-euro coin (that would be a quarter for those of you in the United States or Canada, an Australian dollar, or a British ten pence piece), the situation is getting critical. What you have at this stage is wet snow, which holds a lot of water and is very sticky. Instead of falling through tree branches, wet snow adheres to them, accumulating in thick, heavy layers. Wet snow falling on a tall sturdy tree can break a lot of limbs. It is even worse for adolescent trees. They are standing there with their lanky trunks and small crowns, waiting for it to be their turn to grow. They are either broken by snow loads or bent down so far that they can't right themselves again. Very small trees, however, are not in danger, because their little trunks are simply too short. Pay attention on your next walk out into the forest. Right there among the middle-aged trees, you will find several that have been bent beyond hope of recovery in just such a weather event.

Hoarfrost is similar to snow, but it's much more romantic. At least we think so, because the plants and trees look as though they've been sprinkled with sugar. When below-freezing temperatures and foggy conditions occur together, fine drops of moisture immediately freeze wherever they touch a branch or a needle. After a few hours, the whole forest looks white, even though not a flake of snow has fallen. If the weather conditions persist for days, hundreds of pounds of frosty ice crystals can accumulate in the treetops. When the sun finally breaks through a hole in the fog, all the trees sparkle as though they were in a fairy tale. But unfortunately for them, they are in the real world, and they are groaning under the weight of the ice and beginning to bend dangerously. Woe to the tree that has a weak spot in its wood. Then a dry crack echoes through the forest like a gunshot, and the whole crown comes tumbling down.

In Central Europe, hoarfrost occurs on average every ten years, and this means a tree has to endure up to fifty such events in its lifetime. The less integrated the tree is in a community of its own species, the greater the danger. Loners standing unprotected out in the cold fog succumb markedly more often than well-connected individuals in a dense forest who can lean on their neighbors for support. Moreover, the wind tends to blow over dense forest canopies, so mostly it is just the highest branch tips that get heavily blanketed with ice crystals.

But the weather has still more tricks up its sleeve. There is lightning, for example. An old German saying about storms in the forest, *"Eichen sollst du weichen, Buchen sollst du suchen,"* translates as "Avoid oaks, seek beeches." The saying originates in the fact that on some gnarly old oaks you can see a channel a few inches wide extending down the trunk where a lightning strike has split the bark open and penetrated deep into the wood. I've never seen a scar like this on the trunk of a beech. But the conclusion that lightning never strikes beech trees is as false as it is dangerous. Large old beeches offer no protection from lightning because they are struck just as often. The main reason there is next to no damage on beech trees is because their bark is so smooth.

During a thunderstorm, it rains, and the water that sheets down the wrinkle-free surface of beech bark creates a continuous film. When lightning strikes, the electricity travels down the outside of this film because water conducts electricity much better than wood. Oaks, however, have rough bark. The rainwater that runs down their trunks forms little cascades and drips to the ground in hundreds of mini-waterfalls. Therefore, the flow of electricity from the lightning strike is constantly interrupted. When this happens, the point of least resistance becomes the damp wood of the outer growth rings, which the tree uses to transport water. In response to the energy surge from the lightning strike, the sapwood explodes as though it has been shot, and years later, the scar bears witness to the oak's misfortune.

Douglas firs, which are native to North America but now grow in Central Europe as well, react in much the same way as oaks, but in their case, their roots seem to be super sensitive. In the forest I manage I've observed two lightning strikes where not only the tree that was struck died, but another ten Douglas firs within a radius of 50 feet of the strike experienced the same fate. Clearly, the surrounding trees were connected to the victim underground, and that day, instead of life-giving sugar, what they received was a deadly serving of electricity.

In thunderstorms with a lot of lightning something else can happen—fire. I experienced that once in the middle of the night, when fire trucks rushed into our community forest to extinguish a small fire. Lightning had struck a hollow old spruce. The flames inside the tree were protected from the pouring rain, and they were licking their way up the rotten wood. The fire was put out quickly, but even without help, not much would have happened. The surrounding forest was sopping wet and the fire would very likely not have caught hold in the rest of the stand. Nature doesn't expect fires in native Central European forests. The once-dominant deciduous trees don't catch fire because their wood doesn't contain any resins or essential oils. As a result, none of the trees have developed any mechanism to react to heat. The cork oaks of Portugal and Spain are a testament to the fact that such a mechanism even exists. The cork oaks' thick bark protects them from the heat of grass fires and allows the buds that lie under the bark to start growing again once the fire has passed.

In Central European latitudes, though, the monocultures of plantation spruce and pines can fall prey to fire when the trees' discarded needles get bone dry in summer. But why do conifers store so many flammable substances in their bark and needles anyway? If fires are the order of the day in their native latitudes, wouldn't they be better off if they were highly flame resistant? It wouldn't be possible for a tree like the Swedish spruce in Dalarna, which is more than 9,500 years old, to reach such a ripe old age if it was engulfed by fire every two hundred years. I think it is careless people—the kind of people who leave their campfires unattended, for example—who have been responsible for fiery disturbances in the forest for thousands of years. The small number of lightning strikes that actually start small localized fires are so rare that European tree species never adapted to them. Pay attention to the cause the next time you hear a news report about a forest fire: most are attributed to human activity.

In North America, as in Europe, there have been people around since the last ice age, tinkering with fire. And so it's probable that most forest fires on that continent have also been caused by human activity—a clearing burned for planting subsistence crops here, a carelessly discarded stub of smoldering tobacco there. But Nature has a role to play, too.

Left to their own devices, North American forests experience natural fire cycles. Where the climate is naturally moist and cool, lightning strikes soon fizzle out on the damp forest floor, and forest fires may occur only every few hundred years. In areas where needles and twigs on the ground are often bone dry, lightning can spark fires as often as every couple of years. Fires left to burn through the forest on this natural cycle usually stay at ground level, getting hot enough to burn away brush in the understory and leaving established trees blackened but unscathed.

But just as people spark fires, they also rush to put them out. On the floors of forests not swept by a regular cycle of low-intensity fires, piles of kindling build up, just waiting for a spark. In these conditions, instead of staying low and clearing the understory, fires soon escalate and climb up into the canopy. As the crowns ignite, windblown embers land on neighboring trees, and so the fires spread. Low-level ground fires become raging infernos, leaving acres of blackened slopes in their wake.

Many trees in North America are adapted to natural cycles of ground fires. Ponderosa pines and giant redwoods have evolved thick bark to protect their sensitive cambium. Jack-pines have cones that pop open in heat so that their seeds fall onto a forest floor cleared of vegetation, landing on a soft bed of ash that is a perfect place for life to start anew. However, the character of forest fires in North America has been changed by naturally increasing drought conditions and the human practice of fire suppression, and forests that would once have survived, or even thrived, in the face of fire are now threatened by its destructive force.[64]

Less dangerous but much more painful to trees is a phenomenon that even I didn't know about until recently. The forester's lodge where we live lies on a mountain ridge at an elevation of barely 1,600 feet. The streams all around, which are deeply carved into the landscape, don't do the forest any harm. Quite the opposite. However, large rivers are something completely different. They regularly overflow their banks, and therefore, very specific ecosystems grow on either side of them: forested riparian meadows. Which species of trees get established in these meadows depends on the kind of high water and how often it happens. If the floods are fast flowing and last for many months of the year, then willows and poplars fit the bill. They can cope with long periods in wet conditions. You usually find these conditions close to the river, and this is where willow and poplar meadows get established. Farther away and often a few yards higher up, floods occur less often, and when they do—in spring after the snow melts—then you find large pools of slow-moving water. By the time the trees leaf out, most of the water has already drained away, and in such conditions, oaks and elms feel right at home. In contrast to the areas where willows and poplars grow, these hardwood meadows are very sensitive to summer flooding. In summer floods, otherwise robust trees can die, because their roots suffocate.

Some winters, however, the river really causes the trees pain. On a trip I made through a hardwood meadow in the middle of the Elbe River, I noticed loose strips of bark on all the trees. The damage was all at the same height up the trunks: about 6 feet above ground level. I'd never seen anything like it, and I couldn't figure out what might have caused the damage. The other people on the trip were at as much of a loss as I was, until the staff at the biosphere preserve solved the puzzle: the damage was caused by ice. When the Elbe froze over in particularly cold winters, thick ice floes formed. When the air and the water warmed up in spring, the ice floated between the oaks and elms on the floodwater, bumping up against the tree trunks. As the water was at the same level everywhere on the meadow, the wounds were to be found at the same height on all the trees.

In the context of climate change, one day the movement of ice on the Elbe will be a thing of the past. But the scars of the older trees—at least the ones that have experienced all kinds of capricious weather since the early twentieth century—will bear testimony to these events for a long time to come.

32

— IMMIGRANTS —

THANKS TO THE migration of trees, the forest is constantly changing. And not just the forest—all of Nature. And that's why many human attempts to conserve particular landscapes fail. What we see is always a brief snapshot of a landscape that only seems to be standing still. The illusion is almost perfect in the forest, because trees are among the slowest-moving beings with which we share our world and changes in the natural forest are observable only over the course of many human generations. One of these changes is the arrival of new species.

Thanks to the botanical souvenirs early plant hunters brought back to their homelands and more recent arrivals because of the forest industry, a huge number of tree species have been introduced that would never have found their way to Europe on their own. Names like Douglas fir, Japanese larch, and grand fir don't occur in European folk songs or poems because they have not yet become fixtures in Europe's shared social memory. The process works in the other direction, as well. European arrivals make their own mark when travelers in search of a new life bring memories of home along in their luggage.

Immigrants have a special status in the forest. In contrast to tree species that have migrated naturally, they arrive without their typical ecosystems. In some cases, just their seeds were imported, which means that most of the fungi and all of the insects remained back in their homelands. Douglas Fir & Co. could make a completely new start in Europe. That can certainly have its advantages. There are absolutely no illnesses because of pests—at least not in the first decades. People had a similar experience in Antarctica. The air there is almost completely devoid of germs or dust, which would be ideal for people with allergies, if only the continent were not so isolated. When trees hop over to a new continent with our help, it's like a big breath of fresh air for them. The lucky ones find fungal partners for their roots among the nonspecialists. Beaming with health, the new arrivals grow mighty trunks in European forests, and they do so in very short periods of time. No wonder they seem superior to the native species—at least in some locations.

Trees that migrate under their own steam can establish themselves only where they feel completely at home. Not only the climate but also the type of soil and the moisture levels must fit their lifestyles if they are going to prevail in the presence of the old trees that already rule the forest. For trees that we humans introduce into the forest, the long-term outcome is a bit like a game of roulette. You never know exactly what's going to happen. The black cherry, for example, is a deciduous tree from North America that has a wonderfully beautiful trunk and high-quality wood when it grows there. No question—European foresters wanted to have a tree like that in their forests. But after a few decades, disillusionment set in. In their new land, the trees grew crooked and lopsided and hardly got taller than 65 feet, and they barely grew at all under the pines of eastern and northern Germany. The trees fell out of favor, but by now people couldn't get rid of them. Deer spurned their bitter branches, preferring to nibble away at beeches, oaks, or, if absolutely necessary, even pines. And so the black cherry got the burdensome arboreal competition off its back, and the newcomer keeps expanding its territory.

The Douglas fir can also tell you a tale or two about the uncertainty of the future. In some places, after growing for more than a hundred years, they have become impressive giants. Other forests, however, have had to be cut down in their entirety before they matured, as I experienced firsthand in my intern year in forestry school. A small forest of Douglas firs, barely forty years old, was beginning to die. Scientists puzzled over this for a long time. Whatever could have caused this decline? It wasn't fungi, and insects were ruled out as well. The culprit finally turned out to be an excess of manganese in the soil, which, apparently, the Douglas firs couldn't tolerate.

It also turns out there is no such thing as "the Douglas fir," as separate varieties with completely different characteristics were imported to Europe. Those from the Pacific coast are the best fit. Their seeds, however, got mixed with seeds from inland species that grow a long way from the ocean. And to complicate the situation further, both crossbreed easily, producing offspring, all of which express characteristics that are completely unpredictable. Unfortunately, it often takes at least forty years before you can tell whether the trees are healthy or not. If they are, they keep their vivid blue-green needles and thick crowns with tightly packed branches. The trunks of hybrids that contain too many genes from inland trees begin to bleed resin and their needles look distressed. In the end, this is simply a natural correction, albeit a cruel one. Genetic misfits are discarded, even if the process plays out over many decades.

Our native beeches have had no trouble showing these interlopers the door. They employ the same strategy they use in their competition with oaks. The deciding factor that has allowed beeches to win out over Douglas firs over the course of centuries is their ability to grow in the deepest, darkest shade under large trees. The offspring of the North American mothers need much more light and perish in the kindergartens established by our native deciduous trees. It is only when people lend a helping hand by repeatedly clearing trees so that sunlight reaches the ground that the little Douglas firs stand a chance.

It's dangerous when foreigners pop up that are genetically very similar to native species. The Japanese larch is just such a case. When it arrived here, it met the European larch. The European larch often grows crooked and, in addition, quite slowly, and so in the last century it was often replaced with the Japanese tree. Both species cross easily to form hybrids. This raises the danger that one day, a long time from now, the last purebred European larches will disappear. There's just such a mixing and muddling of genes going on in the forest I manage in the Eifel mountains, where neither species is native. Another candidate for extinction is the black poplar, which mixes with cultivated hybrid poplars that have been crossed with Canadian poplars.

But most introduced species pose no threat to native trees. Without our help, a number of them would have disappeared again after a couple of hundred years at the most. Even with our help, the survival of the new arrivals is questionable in the long term. For the pests that plague them take advantage of global trade. It is true that there is no active import of these organisms—after all, who would want to introduce damaging pests? Yet, slowly but surely, fungi and insects are making their way across the Atlantic or the Pacific in imported lumber and establishing themselves in Europe. Often they come in packing materials, such as wood pallets that haven't been heated to sufficiently high temperatures to kill harmful organisms. And parcels sent by private individuals from overseas sometimes contain living insects. I have personal experience of this. I had an antique moccasin from North America shipped to my home in Germany. As I unpacked the leather footwear from its newspaper wrapping, a number of small brown beetles crawled in my direction. I caught them as quickly as I could, squished them, and disposed of them in the trash. Squishing bugs might sound odd coming from the pen of a conservationist, but introduced insects, once they get established, are life threatening not only for introduced species but also for natives.

The Asian long-horned beetle poses just such a threat. It probably traveled to Europe and other parts of the world from China in packing crates. The beetle is an inch long and has 2-inch-long antennae. To us, it's a beautiful-looking beetle. Its dark body is flecked with white, and it has black and white bands on its legs and antennae. Deciduous trees, however, find it decidedly less attractive, because it lays its eggs individually in numerous small splits in their bark. Voracious larvae hatch and feed, and adult beetles drill thumb-sized exit holes in the trunk. These holes are then attacked by fungi, and eventually the trunk breaks. In Europe, the beetles are still concentrated in urban areas, making life even more difficult for the "street kids." We don't yet know if they will spread to forested areas away from urban settlements, because the beetles are lazy and prefer to stay within a radius of a few hundred yards of the place where they were born.

Another import from Asia behaves very differently. This particular fungus, ash dieback fungus, is well on its way to finishing off most of the ash trees in Europe. Its fruiting bodies look harmless, even rather cute. They are just teeny-weeny mushrooms that grow on the stalks of fallen leaves. The fungal filaments themselves, however, run amok in the trees and kill one ash after another. A few ash trees seem to survive the repeated assaults, but it is questionable whether there will be ash forests lining the banks of European streams and rivers in the future. In connection with this, I sometimes wonder if foresters don't play a role in the spread of the disease. I visited damaged forests in southern Germany, and then afterward, I was out and about in the forest I manage—wearing the same shoes! Might there have been tiny fungal spores on my soles that traveled into the Eifel mountains as stowaways? Whatever the case may be, since then, the first ash trees in Hümmel have also been struck with the disease.

Despite all this, I am not anxious when I think about the future of our forests. For on large continents (and the Eurasian continent is the largest one of all) species have to come to grips with new arrivals all the time. Migrating birds bring small animals, fungal spores, or the seeds of new species of trees tucked in their feathers, or these organisms are blown in by turbulent storms. A five-hundred-year-old tree has surely had a few surprises in its life. And thanks to the great genetic diversity in a single species of tree, there is always a sufficient number of individuals that can rise to a new challenge.

If you live in or have traveled to Germany, you might well have already noticed some of the new "naturalized" avian citizens that have turned up without any help from people. Perhaps the Eurasian collared dove, which arrived in Germany from the Mediterranean in the 1930s. Then there is the fieldfare, a type of thrush. This gray-brown bird with dark speckles has been migrating westward for two hundred years. It started in the northeast and has now reached France. We don't yet know what surprises these birds might have brought with them in their feathers.

A decisive factor in how robust native forests are in the face of such changes is how unspoiled they are. The more intact the social connections and the more moderated the microclimate under the trees, the more difficult it is for foreign invaders to get established. Plants that make headlines are classic examples of this. Take giant hogweed (also known as wild parsnip or wild rhubarb). It originally came from the Caucasus and grows more than 10 feet tall. The white flower heads can measure up to 18 inches across, and because they are so pretty, the plant was imported into Central Europe and elsewhere in the nineteenth century. The plants escaped out of the gardens where they were planted and since then have been spreading across the countryside with ease.

Giant hogweed is considered extremely dangerous because its sap, in combination with ultraviolet light, can burn human skin. Every year, millions are spent digging up plants and destroying them, without any great success. However, hogweed can spread only because the original forested meadows along the banks of rivers and streams no longer exist. If these forests were to return, it would be so dark under the forest canopy that hogweed would disappear. The same goes for Himalayan balsam and Japanese knotweed, which also grow on the riverbanks in the absence of the forests. Trees could solve the problem if people trying to improve things would only allow them to take over.

I have written so much about nonnative species that this might be the place to address the question of what the term "native" means. We tend to call species native if they occur naturally within a country's borders. A classic example from the animal world is the wolf, which reappeared in most countries in Central Europe in the 1990s and since then has been considered a permanent part of the fauna. It was found in Italy,

France, and Poland much earlier than that. This means that the wolf has been native to Europe for a long time, just not in each individual country. But isn't even this geographic unit too broad? When we talk about porpoises being native to Germany, does that mean they also make their home in the upper reaches of the Rhine? As you can see, that definition wouldn't make any sense. Native must be understood on a much smaller scale and be based not on human borders but on habitats.

Habitats are defined by their features (water, terrain, topography) and by the local climate. After the last ice age, trees moved into habitats where they found conditions that suited them. That means, for example, that spruce occur naturally (and, therefore, can be considered native) at an elevation of 4,000 feet in the Bavarian Forest, but they do not occur naturally (and, therefore, cannot be considered native) 1,300 feet lower and only half a mile away, where beeches and firs hold sway. Specialists have come up with the term "habitat specific," which simply means each species has habitats where they are happy to grow. In contrast to our large-scale country borders, habitat borders for species are like a proliferation of small states. When people ignore these boundaries and bring spruce and pines down to warmer elevations, these conifers are not natives in this new location; they are immigrants. And with that we have arrived at my favorite example: red wood ants.

In Europe, red wood ants are icons of nature conservation. In many locations they are mapped, protected, and in cases of conflict, resettled. There can be no objection to this because what we are talking about here is a threatened species.

Threatened? And yet red wood ants are immigrants, too, and therefore, I would argue that no special efforts are necessary for their protection. They travel on the coattails of commercially grown spruce and pines. You could say they hang on to the needles for dear life, for without the conifers' spiny, narrow needles, they can't build their anthills. And this proves that they were not present in the original native deciduous forests. Moreover, they love the sun, and they need it to shine on their nests for at least a few hours a day. Especially in spring and fall, when it is bitterly cold, a few warm rays ensure additional days when the ants can rummage around. Dark beech woods are, therefore, ruled out as habitat, and red wood ants are forever thankful to foresters for planting huge expanses of spruce and pines.

33

— HEALTHY FOREST AIR —

FOREST AIR IS the epitome of healthy air. People who want to take a deep breath of fresh air or engage in physical activity in a particularly agreeable atmosphere step out into the forest. There's every reason to do so. The air truly is considerably cleaner under the trees, because the trees act as huge air filters. Their leaves and needles hang in a steady breeze, catching large and small particles as they float by. Per year and square mile this can amount to 20,000 tons of material.[65] Trees trap so much because their canopy presents such a large surface area. In comparison with a meadow of a similar size, the surface area of the forest is hundreds of times larger, mostly because of the size difference between trees and grass. The filtered particles contain not only pollutants such as soot but also pollen and dust blown up from the ground. It is the filtered particles from human activity, however, that are particularly harmful. Acids, toxic hydrocarbons, and nitrogen compounds accumulate in the trees like fat in the filter of an exhaust fan above a kitchen stove. But not only do trees filter materials out of the air, they also pump substances into it. They exchange scent-mails and, of course, pump out phytoncides, both of which I have already mentioned.

Forests differ a great deal from one another depending on the species of trees they contain. Coniferous forests noticeably reduce the number of germs in the air, which feels particularly good to people who suffer from allergies. However, reforestation programs introduce spruce and pines to areas where they are not native, and the newcomers experience substantial problems in their new habitats. Usually, they are brought to low elevations that are too warm and dry for conifers to thrive. As a result, the air is dustier, as you can clearly see when the dust motes are backlit by sun streaming down on a summer's day. And because the spruce and pines are constantly in danger of dying of thirst, they are easy prey for bark beetles, which come along to make a meal of them. At this point, frantic scent-mails begin to swirl around in the canopy. The trees are "screaming" for help and activating their arsenal of chemical defenses. You absorb all of this with every breath of forest air you take into your lungs. Is it possible that you could unconsciously register the trees' state of alarm?

Consider this. Threatened forests are inherently unstable, and therefore, they are not appropriate places for human beings to live. And because our Stone Age ancestors were always on the lookout for ideal places to set up camp, it would make sense if we could intuitively pick up on the state of our surroundings. There is a scientific observation that speaks to this: the blood pressure of forest visitors rises when they are under conifers, whereas it calms down and falls in stands of oaks.[66] Why don't you take the test for yourself and see in what type of forest you feel most comfortable?

Whether we can somehow listen in on tree talk is a subject that was recently addressed in the specialized literature.[67] Korean scientists have been tracking older women as they walk through forests and urban areas. The result? When the women were walking in the forest, their blood pressure, their lung capacity, and the elasticity of their arteries improved, whereas an excursion into town showed none of these changes. It's possible that phytoncides have a beneficial effect on our immune systems as well as the trees' health, because they kill germs. Personally, however, I think the swirling cocktail of tree talk is the reason we enjoy being out in the forest so much. At least when we are out in undisturbed forests.

Walkers who visit one of the ancient deciduous preserves in the forest I manage always report that their heart feels lighter and they feel right at home. If they walk instead through coniferous forests, which in Central Europe are mostly planted and are, therefore, more fragile, artificial places, they don't experience such feelings. Possibly it's because in ancient beech forests, fewer "alarm calls" go out, and therefore, most messages exchanged between trees are contented ones, and these messages reach our brains as well, via our noses. I am convinced that we intuitively register the forest's health. Why don't you give it a try?

Contrary to popular opinion, the air in the forest is not always particularly rich in oxygen. This essential gas is released when water and carbon dioxide are broken down during photosynthesis. Every day in summer, trees release about 29 tons of oxygen into the air per square mile of forest. A person breathes in nearly 2 pounds of oxygen a day, so that's the daily requirement for about ten thousand people. Every walk in the forest is like taking a shower in oxygen. But only during the day. Trees manufacture large amounts of carbohydrates not only to lay them down as wood but also to satisfy their hunger. Trees use carbohydrates as fuel, just as we do, and when they do, they convert sugar into energy and carbon dioxide. During the day, this doesn't affect the air much because after all the additions and subtractions, there is still that surplus oxygen I just mentioned. At night, however, the trees don't photosynthesize, and so they don't break down carbon dioxide. Quite the opposite, in fact. In the darkness, it's all about using carbohydrates, burning sugar in the cells' power-generating stations, and releasing carbon dioxide. But don't worry, you won't suffocate if you take a nighttime ramble! A steady movement of air through the forest ensures that all the gases are well mixed at all times, and so the drop in oxygen near the ground is not particularly noticeable.

How does a tree breathe anyway? You can see a part of its "lungs." These are the needles or leaves. They have narrow slits on their undersides that look a bit like tiny mouths. The tree uses these openings to exhale oxygen and breathe in carbon dioxide. At night, when the tree is not photosynthesizing, it does the reverse. It's a long way from the leaves, down the trunk, to the roots, and that's why tree roots can breathe as well. If they didn't, deciduous trees would die in winter when they discard their aboveground lungs. But the trees keep ticking over and their roots even grow a little, so energy must be produced with the help of the trees' reserves, and for this the trees need oxygen. And that is why it is so awful for a tree if the soil around its trunk has been so compacted that the small air pockets in the soil have been crushed. The tree's roots suffocate, or at least have difficulty breathing, with the result that the tree gets sick.

But let's get back to breathing at night. It's not only the trees that are exhaling large amounts of carbon dioxide in the dark. In leaves, in dead wood, and in other rotting plant material, microscopic creatures, fungi, and bacteria are busy in a round-the-clock feeding frenzy, digesting everything edible and then excreting it as humus. In winter, the situation gets even tougher. This, of course, is when the trees are hibernating and even during the day, the oxygen levels are not being topped up, while the soil organisms continue merrily working away underground. They generate so much heat that even in a hard frost, the ground doesn't freeze down more than 2 inches. Does that mean that a forest in winter is dangerous? What saves us is the global circulation of air, which constantly blows fresh marine air over the continent. A multitude of algae live in salt water. Thanks to them, large amounts of oxygen bubble up out of the ocean year round. Algal activity in the oceans balances the oxygen deficit in Central European forests in the winter so well that we can breathe deeply even when we are standing under beeches and spruce covered in snow.

On the subject of sleep: have you ever considered whether this is something trees even need? What would happen if we wanted to help them, and so we provided them with light at night as well as during the day so that they could manufacture more sugar? According to current research, that would be a bad idea. It seems trees need their rest just as much as we do, and sleep deprivation is as detrimental to trees as it is to us.

In 1981, the German journal *Gartenamt* reported that 4 percent of oak deaths in one American city happened because the trees were subjected to light every night. And what about the long period of hibernation? This has already been tested unwittingly by some forest fans. I wrote about this in chapter 22, "Hibernation." They brought young oaks and beeches into their houses, where they kept them in pots on windowsills. In cozy living rooms there's no such thing as winter as far as the temperature is concerned, which means most of the young trees couldn't take a breather and just continued to grow. But at some point, lack of sleep exerted its revenge and the plants, which had seemed so full of life, died. Now you could argue that some winters aren't really very wintery, and at least at lower elevations, there are hardly any frosty days. Despite this, deciduous trees still lose their leaves and don't grow them again until spring, because, as I have already mentioned, they also measure day length. But isn't that also the case for the little trees on the windowsill? It might be the case if the heating were turned off and winter evenings were spent in the dark, but hardly any of us are willing to give up the comfortable temperatures (around 70 degrees Fahrenheit) and warm white electric light that conjure up artificial summers inside our houses. And no Central European forest tree can endure eternal summer.

34

— WHY IS THE FOREST GREEN? —

WHY DO WE find it so much more difficult to understand plants than animals? It's because of the history of evolution, which split us off from vegetation very early on. All our senses developed differently, and so we have to use our imaginations to get even the slightest idea of what is going on inside trees. Our color vision is a good example. I love the combination of a bright-blue sky over a canopy of lush green. For me, this color combination is Nature at its most idyllic and the most relaxing color combination I can imagine. Would trees agree with me? Their answer would probably be: "More or less."

Beeches, spruce, and other species certainly find blue sky, which means lots of sun, equally agreeable. For them, however, the color isn't so much romantic or moving as it is a flag that signals, "The buffet is open." For a cloudless firmament means high-intensity light and, therefore, optimal conditions for photosynthesis. Frantic activity for maximum output is the order of the day. Blue means a lot of work. The trees get full as they convert light, carbon dioxide, and water into supplies of sugar, cellulose, and other carbohydrates.

Green, however, has a completely different significance. Before we get to the typical color of most plants, we first have to answer another question: Why is the world full of color anyway? Sunlight is white, and when it is reflected, it is still white. And so we should be surrounded by a clinical-looking, optically pure landscape. That this is not what we see is because every material absorbs light differently or converts it into other kinds of radiation. Only the wavelengths that remain are refracted and reach our eyes. Therefore, the color of organisms and objects is dictated by the color of the reflected light. And in the case of leaves on trees, this color is green.

But why don't we see leaves as black? Why don't they absorb all the light? Chlorophyll helps leaves process light. If trees processed light super-efficiently, there would be hardly any left over—and the forest would then look as dark during the day as it does at night. Chlorophyll, however, has one disadvantage. It has a so-called green gap, and because it cannot use this part of the color spectrum, it has to reflect it back unused. This weak spot means that we can see this photosynthetic leftover, and that's why almost all plants look deep green to us. What we are really seeing is waste light, the rejected part that trees cannot use. Beautiful for us; useless for the trees. Nature that we find pleasing because it reflects trash? Whether trees feel the same way about this I don't know, but one thing is for certain: hungry beeches and spruce are as happy to see blue sky as I am.

The color gap in chlorophyll is also responsible for another phenomenon: green shadows. If beeches allow no more than 3 percent of sunlight to reach the forest floor, it should be almost dark down there during the day. But it isn't, as you can see for yourself when you take a walk in the forest. Yet hardly any other plants grow here. The reason is that shadows are not all the same color. Although many shades of color are filtered out in the forest canopy—for example, very little red and blue make their way through—this is not the case for the "trash" color green. Because the trees can't use it, some of it reaches the ground. Therefore, the forest is transfused with a subdued green light that just happens to have a relaxing effect on the human psyche.

In my garden, a single beech seems to prefer red leaves. It was planted by one of my predecessors, and it has grown into a large tree. I don't like it very much because, in my opinion, the leaves look unhealthy. You can find trees with reddish leaves in many parks, where they are supposed to inject interest in an otherwise monotonous sea of green. The common English name for my tree is copper beech. (In German it is known as a *Blutbuche*, or "blood beech," which doesn't make me any more inclined to like it.) But really, I think I feel sorry for this tree because its deviation from the traditional appearance of a beech works to its disadvantage.

The color is the result of a metabolic disorder. Young developing leaves on normal trees are often tinged red thanks to a kind of sun block in their delicate tissue. This is anthocyanin,

which blocks ultraviolet rays to protect the little leaves. As the leaves grow, the anthocyanin is broken down with the help of an enzyme. A few beeches or maples deviate from the norm because they lack this enzyme. They cannot get rid of the red color, and they retain it even in their mature leaves. Therefore, their leaves strongly reflect red light and waste a considerable portion of the light's energy. Of course, they still have the blue tones in the spectrum for photosynthesis, but they are not achieving the same levels of photosynthesis as their green-leaved relatives. These red trees keep appearing in Nature, but they never get established and always disappear again. Humans, however, love anything that is different, and so we seek out red varieties and propagate them. One man's trash is another man's treasure is one way to describe this behavior, which might stop if people knew more about the trees' circumstances.

The main reason we misunderstand trees, however, is that they are so incredibly slow. Their childhood and youth last ten times as long as ours. Their complete life-span is at least five times as long as ours. Active movements such as unfurling leaves or growing new shoots take weeks or even months. And so it seems to us that trees are static beings, only slightly more active than rocks. And the sounds that make the forest seem so alive—the rustling of the crowns in the wind, the creaking of branches and trunks as they blow back and forth—are only passive swaying motions that are, at best, a nuisance for the tree. It's hardly any wonder that many people today see trees as nothing more than objects. At the same time, some of the processes under the trunk happen much more quickly than the ones we can see. For instance, water and nutrients—that is to say, "tree blood"—flow from the roots up to the leaves at the rate of a third of an inch per second.[68]

Even conservationists and many foresters are victims of optical illusions in the forest. This is hardly surprising. People rely heavily on sight, and so we are particularly influenced by this sense. Thus, ancient forests in Central European latitudes often strike us as being dull and species poor when we see them for the first time. The diversity of animal life plays out mostly in the microscopic realm, hidden from the eyes of forest visitors. We notice only the larger species, such as birds or mammals, and we don't see them very often because typical forest dwellers are mostly quiet and very shy. And so when I take people visiting my forest around the old beech preserves, they often ask why they hear so few birds. Species that live out in the open often make more noise and take less trouble to keep out of sight. Perhaps you are familiar with this behavior from your own garden, where tits and chickadees, blackbirds and robins quickly get used to you and don't bother to hop or fly away more than a few yards when you come along. Even the butterflies in the forest are mostly brown and gray and blend in with bark when they land on a tree trunk, whereas those that fly in wide-open spaces vie with one another in such a symphony of color and iridescence that it's almost impossible to miss them. It's the same with the plants. Forest species are mostly small and look very much alike. There are so many hundreds of species of mosses, all tiny, that even I have lost track, and the same goes for the diversity of lichens. How much more attractive are the plants of the open plains.

The radiant foxglove towering up to 6 feet tall, yellow ragwort, the sky-blue forget-me-not—such splendor brightens the hiker's heart.

It's no wonder that some conservationists are thrilled when storms or commercial forestry operations disturb the forest ecosystem by opening up large clearings. They truly believe the open space increases species diversity, and they miss the fact that this is traumatic for the forest. In exchange for a few species adapted to open areas that now feel like a million dollars basking happily in the bright sun, hundreds of microscopic organisms of little interest to most people die out locally. A scientific study by the Ecological Society of Germany, Austria, and Switzerland concluded that although increased forest management leads to increased richness in the diversity of plant life, this is no cause for celebration but rather proof of the level of disturbance of the natural ecosystem.[69]

35

— SET FREE —

IN THESE TIMES of dramatic environmental upheaval, our yearning for undisturbed nature is increasing. Countries around the world are enacting legislation to protect what remains of their original forests. In the United Kingdom, the designation "ancient woodlands" affords some protection to woodlands that have existed continuously since at least the 1600s. Often formerly the property of large estates, over their history they have been intensively managed for wood and wildlife, and so, although the wood itself may be ancient, the trees that grow there may not. In Australia, the term "old-growth forest" helps protect some ancient forests from logging, but as economic interests push back, arguments are inevitably raised about the precise meaning of the term.

In the United States, forest preserves, such as the Adirondack and Catskill parks in New York State, keep economic interests out of the forests. According to the state constitution, the preserve "shall be forever kept as wild forest lands," and the timber shall not be "sold, removed or destroyed." In the wilderness areas of these preserves, most structures are not allowed, power vehicles are banned, and chainsaws require special permits. What started as a measure to ensure that excessive logging in the nineteenth century didn't lead to soil erosion and silting up of the economically important Erie Canal has turned into a resource dedicated to the forest itself and visitors who "leave no trace" as they pass through.

Even more remote is the Great Bear Rainforest in northern British Columbia, which covers almost 25,000 square miles along the rugged coast. Half of this area is forested, including about 8,900 square miles of old-growth trees. This primeval forest is home to the rare spirit bear, which although it is white, is not a polar bear but a black bear with white fur. First Nations in the area have been fighting since the 1990s to protect their homelands. On February 1, 2016, an agreement was announced to keep 85 percent of the forest unlogged, though it does allow for 15 percent of the trees, mostly old growth at low elevations, to be removed. After a long hard struggle, some progress, at least, has been made in protecting this very special place. Chief Marilyn Slett, president of Coastal First Nations, is well aware of the forest's importance: "Our leaders understand our well being is connected to the well being of our lands and waters... If we use our knowledge and our wisdom to look after [them], they will look after us into the future."[70] The Kichwa of Sarayaku, Ecuador, see their forest as "the most exalted expression of life itself."[71]

In densely populated Central Europe, the forest is the last refuge for people who want to let their spirits soar in landscapes untouched by human hand. But there really isn't any undisturbed nature left here. The old-growth forests disappeared centuries ago, first to the axes and finally under the plows of our forefathers, who were beset by famine. It's true that today, once again, there are large tree-covered areas next to settlements and fields, but these are plantations rather than forests—the trees are all the same species and the same age. Politicians are beginning to debate whether such plantings can really be called forests at all.

There's consensus among German politicians that 5 percent of the forests should be left to their own devices so that they can become the old-growth forests of tomorrow. At first, that doesn't sound like much, and it's downright embarrassing when compared with states in tropical parts of the world, the ones we always reproach for the lack of protection for their rain forests. But at least it's a start. Even if only 2 percent of the forests in Germany were freed from human interference, that's still more than 770 square miles. You could observe the free play of natural forces in such areas. In contrast to nature preserves, which are always carefully groomed, what would be preserved here would be doing absolutely nothing. In scientific terms, this is known as "process conservation." And because Nature is completely uninterested in what we humans want, the processes don't always progress as we would like them to.

Basically, the more severely out of balance the protected area, the more intense the process of returning to undisturbed forest. The most extreme contrast would be a bare field, followed by a home lawn that is mowed every week. I notice this around our forest lodge, too. There are always oak, beech, and birch seedlings popping up in the grass. If I didn't cut them off regularly, within five years I'd have a stand of young trees about 6 feet tall, and our little piece of paradise would disappear behind their foliage.

In forested areas of Central Europe, it is the return of spruce and pine plantations to ancient forest that is most dramatic. And it is precisely these forests that are often part of newly established national parks, because people usually don't want to consolidate them with the ecologically more valuable deciduous forests. It doesn't really matter. The future old-growth forest is just as happy to develop from a monoculture. As long as people don't meddle, the first drastic changes can be seen after just a few years. Usually, it's the arrival of insects, such as tiny bark beetles, which can now proliferate and spread without hindrance. The conifers were originally planted in symmetrical rows in places that were too warm and dry for them. In these conditions, they are unable to defend themselves from their attackers, and within just a few weeks, their bark is completely dead as a result of the beetles' depredations.

The insect invasion spreads like wildfire through the former commercial forest, leaving in its wake a seemingly dead, barren landscape, strewn with the pale ribs of trees. This bleak scene makes the hearts of the resident sawmill workers bleed, as they would have preferred to put the trunks to good use. They also argue that the devastating sight means tourism can't really get going either. That's understandable if visitors come unprepared. They are expecting to take a walk in what is supposedly an intact forest, and instead of seeing healthy green growth, they encounter a series of hillsides completely covered in dead trees. In the Bavarian National Forest alone, more than 20 square miles of spruce forest have died since 1995—about one quarter of the total area of the park.[72] Dead trunks are clearly more difficult for some visitors to bear than bald, empty spaces.

Most national parks give in to the clamor of complaint and sell to sawmills the trees they have felled and removed from the forest to combat bark beetle infestations. This is a grave mistake. For the dead spruce and pines are midwives to the new deciduous forest. They store water in their dead trunks, which help cool the hot summer air to a bearable temperature. When they fall over, the impenetrable barricade of trunks acts as a natural fence through which no deer can pass. Protected in this way, the small oaks, bird cherries, and beeches can grow up unbrowsed. And when one day the dead conifers rot, they create valuable humus.

But you don't have an established forest yet, because the young trees don't have any parents. There's no one there to slow the growth of the little ones, to protect them, or in case of emergency, to feed them sugar. The first natural generation of trees in a national park, therefore, grows up more or less like the "street kids." Even the mix of tree species is unnatural at first. The former coniferous plantation trees sow their seeds heavily before they depart, so spruce, pines, and Douglas firs grow along with the beeches, oaks, and silver firs. It's at this point that officials usually get impatient. No question, if the conifers that have now fallen into disfavor were to be removed, the future old-growth forest would develop a bit more quickly.

But once you understand that the first generation of trees is going to grow too quickly anyway and, therefore, is not going to get very old—and that the stable social structure of the forest is not going to be laid down until much later—then you can take a more relaxed view.

The plantation trees growing in the mix will depart in less than a hundred years because they will grow above the tops of the deciduous trees and stand unprotected in the path of storms that will ruthlessly uproot them. These first gaps will be vanquished by the second generation of deciduous trees, which can now grow up protected by the leafy canopy formed by their parents. Even if these parents themselves don't grow very old, they will still grow old enough to give their children a slow start. Once these youngsters reach the age of retirement, the future old-growth forest will have achieved equilibrium, and from then on, it will hardly change at all.

It takes five hundred years from the time a national park is established to get to this point. Had large areas of an old deciduous forest that had seen only modest commercial use been put under protection, it would take only two hundred years to reach this stage. However, because all over Germany the forests chosen for protection are forests that are far from their natural state, you have to allow a little more time (from the trees' point of view) and a particularly intense restructuring phase for the first few decades.

There's a common misconception about the appearance of old-growth forests in Europe, should they come to pass. Laypeople often assume that shrubby growth will take over the landscape and forests will become impenetrable. Where today the forests that predominate are at least partially accessible, tomorrow chaos will rule. Forest preserves untouched by foresters for more than a hundred years prove the opposite. Because of the deep shade, wild flowers and shrubs don't have a chance, so the color brown (from old leaves) predominates on the natural forest floor. The small trees grow extremely slowly and very straight, and their side branches are short and narrow. The old mother trees dominate, and their flawless trunks stretch to the sky like the columns in a cathedral.

In contrast to this, there is much more light in managed forests, because trees are constantly being removed. Grass and bushes grow in the gaps, and tangles of brambles prevent detours off the beaten path. When trees are felled and their crowns are left lying on the ground, the debris creates further obstacles. The whole forest presents a troubled and downright messy picture. Old-growth forests, however, are basically very accessible. There are just a few thick dead trunks lying on the ground here and there, which offer natural resting spots. Because the trees grow to be so old, few dead trees fall. Other than that, nothing much happens. Few changes are noticeable in a person's lifetime. Preserves where managed forests are allowed to develop into old-growth forests have a calming effect on Nature and offer better experiences for people seeking rest and relaxation.

And what about personal safety? Don't we read every month about the dangers of walking out under old trees? Falling branches or complete trees that fall across footpaths, sheds, or parked cars? Certainly, that could happen. But the dangers of managed forests are much higher. More than 90 percent of storm damage happens to conifers growing in unstable plantations that fall over with wind gusts of 60 miles an hour. I don't know of a single case where an old deciduous forest left to its own devices for many years has suffered comparable damage in similar weather. And so all I can say is: let's have a bolder approach to wilderness!

36

— MORE THAN JUST — A COMMODITY

If you look at the shared history of people and animals, the final decades of the twentieth century and the first decade or so of the twenty-first century have been positive. It's true there are still factory farms, experiments done on animals, and other ruthless forms of exploitation; however, as we credit our animal colleagues with increasingly complex emotional lives, we are extending rights to them, as well. In Germany, a law that improved animal rights under civil law (referred to in Germany by the shorthand *TierVerbG*) came into force in 1990. The goal of this legislation is to ensure that animals are no longer treated as objects. More and more people are giving up meat altogether or giving more thought to how they buy meat to promote the humane treatment of animals.

I applaud these changes because we are now discovering that animals share many human emotions. And not just mammals, which are closely related to us, but even insects such as fruit flies. Researchers in California have discovered that even these tiny creatures might dream.[73] Sympathy for flies? That's quite a stretch for most people, and the emotional path to the forest is even more of a stretch. Indeed, the conceptual gap between flies and trees is well-nigh unbridgeable for most of us. Large plants do not have brains, they move very slowly, their interests are completely different from ours, and they live their daily lives at an incredibly slow pace. It's no wonder that even though every schoolchild knows trees are living beings, they also know they are categorized as objects.

When the logs in the fireplace crackle merrily, the corpse of a beech or oak is going up in flames. The paper in the book you are holding in your hands right now is made from the shavings of spruce, and birches were expressly felled (that is to say, killed) for this purpose. Does that sound over the top? I don't think so. For if we keep in mind all we have learned in the previous chapters, parallels can definitely be drawn to pigs and pork. Not to put too fine a point on it, we use living things killed for our purposes. Does that make our behavior reprehensible? Not necessarily. After all, we are also part of Nature, and we are made in such a way that we can survive only with the help of organic substances from other species. We share this necessity with all other animals. The real question is whether we help ourselves only to what we need from the forest ecosystem, and—analogous to our treatment of animals—whether we spare the trees unnecessary suffering when we do this.

That means it is okay to use wood as long as trees are allowed to live in a way that is appropriate to their species. And that means that they should be allowed to fulfill their social needs, to grow in a true forest environment on undisturbed ground, and to pass their knowledge on to the next generation. And at least some of them should be allowed to grow old with dignity and finally die a natural death.

What organic farms are to agriculture, continuous cover forests with careful selective cutting are to silviculture. In these forests (called *Plenterwälder* in German), trees of different ages and sizes are mixed together so that tree children can grow up under their mothers. Occasionally, a tree is harvested with care and removed using horses. And so that old trees can fulfill their destinies, 5 to 10 percent of the area is completely protected. Lumber from forests with such species-appropriate tree management can be used with no qualms of conscience. Unfortunately, 95 percent of the current forest practice in Central Europe looks quite different, with the use of heavy machinery and plantation monocultures.

Laypeople often intuitively grasp the need for a change in forest management practices better than forestry professionals do. The public is getting increasingly involved in the management of community forests, and they are insisting the authorities embrace higher environmental standards. We have the example of "forest-friendly" Königsdorf near Cologne, which reached a mediated agreement with the forest service and the regional ministry for natural resources and the environment that heavy machinery no longer be used and deciduous trees of a great age never be cut down.[74] On the other side of the Atlantic, in Virginia, the mission of the nonprofit Healing Harvest Forest Foundation is to "address human need for forest products while creating a nurturing co-existence between the forest and human community." The foundation supports community-based forestry initiatives and promotes the use of horses, mules, and oxen to remove felled trees and the practice of removing single trees that are struggling when harvesting timber, leaving the healthiest standing.[75]

In the case of Switzerland, a whole country is concerned with the species-appropriate treatment of all things green. The constitution reads, in part, that "account [is] to be taken of the dignity of creation when handling animals, plants and other organisms." So it's probably not a good idea to decapitate flowers along the highway in Switzerland without good reason. Although this point of view has elicited a lot of head shaking in the international community, I, for one, welcome breaking down the moral barriers between animals and plants. When the capabilities of vegetative beings become known, and their emotional lives and needs are recognized, then the way we treat plants will gradually change, as well. Forests are not first and foremost lumber factories and warehouses for raw material, and only secondarily complex habitats for thousands of species, which is the way modern forestry currently treats them. Completely the opposite, in fact.

Wherever forests can develop in a species-appropriate manner, they offer particularly beneficial functions that are legally placed above lumber production in many forest laws. I am talking about respite and recovery. Current discussions

between environmental groups and forest users, together with the first encouraging results—such as the forest in Königsdorf—give hope that in the future forests will continue to live out their hidden lives, and our descendants will still have the opportunity to walk through the trees in wonder. This is what this ecosystem achieves: the fullness of life with tens of thousands of species interwoven and interdependent.

And just how important this interconnected global network of forests is to other areas of Nature is made clear by this little story from Japan. Katsuhiko Matsunaga, a marine chemist at the Hokkaido University, discovered that leaves falling into streams and rivers leach acids into the ocean that stimulate the growth of plankton, the first and most important building block in the food chain. More fish because of the forest? The researcher encouraged the planting of more trees in coastal areas, which did, in fact, lead to higher yields for fisheries and oyster growers.[76]

But we shouldn't be concerned about trees purely for material reasons, we should also care about them because of the little puzzles and wonders they present us with. Under the canopy of the trees, daily dramas and moving love stories are played out. Here is the last remaining piece of Nature, right on our doorstep, where adventures are to be experienced and secrets discovered. And who knows, perhaps one day the language of trees will eventually be deciphered, giving us the raw material for further amazing stories. Until then, when you take your next walk in the forest, give free rein to your imagination—in many cases, what you imagine is not so far removed from reality, after all!

NOTE FROM A FOREST SCIENTIST

THE UNDERGROUND SOCIAL networks of trees that Peter Wohlleben describes in his home woodlands of Germany were discovered in the inland temperate rainforests of western North America. In the early 1990s, when searching for clues to the remarkable fertility of these Pacific forests, we unearthed a constellation of fungi linking manifold tree species. The mycelial web, as we later discovered, was integral to the life of the forest. Peter's account that these networks, as in our old-growth forests, are also important to the well-being of the beech, oak, and planted spruce forests of Europe is heartening.

My own search for this web in my home forests began as a quest to understand why weeding paper birches from clear-cut plantations went hand in hand with the decline of planted Douglas firs. In the rows of saplings, I would often see clusters of firs suffering from the loss of their birch neighbors. Yes, trees decline and die naturally—gracefully, beautifully, generously—as an essential part of the irrepressible life cycle of the forest. But this pattern of premature death had been concerning me for some time. The loss of synergy between broad-leaved trees and conifers, it turns out, was a concern of Peter's, too. Across the forests of Europe, planting and weeding to create clean rows has been practiced for centuries.

With the web uncovered, the intricacies of the belowground alliance still remained a mystery to me, until I started my doctoral research in 1992. Paper birches, with their lush leaves and gossamer bark, seemed to be feeding the soil and helping their coniferous neighbors. But how? In pulling back the forest floor using microscopic and genetic tools, I discovered that the vast belowground mycelial network was a bustling community of mycorrhizal fungal species. These fungi are mutualistic. They connect the trees with the soil in a market exchange of carbon and nutrients and link the roots of paper birches and Douglas firs in a busy, cooperative Internet. When the interwoven birches and firs were spiked with stable and radioactive isotopes, I could see, using mass spectrometers and scintillation counters, carbon being transmitted back and forth between the trees, like neurotransmitters firing in our own neural networks. The trees were communicating through the web!

I was staggered to discover that Douglas firs were receiving more photosynthetic carbon from paper birches than they were transmitting, especially when the firs were in the shade of their leafy neighbors. This helped explain the synergy of the pair's relationship. The birches, it turns out, were spurring the growth of the firs, like carers in human social networks. Looking further, we discovered that the exchange between the two tree species was dynamic: each took different turns as "mother," depending on the season. And so, they forged their duality into a oneness, making a forest. This discovery was published by *Nature* in 1997 and called the "wood wide web."

The research has continued unabated ever since, undertaken by students, postdoctoral researchers, and other scientists, with a myriad of discoveries about belowground communication among trees. We have used new scientific tools, as they are invented, along with our curiosity and dreams, to peer into the dark world of the soil and illuminate the social network of trees. The wood wide web has been mapped, traced, monitored, and coaxed to reveal the beautiful structures and finely adapted languages of the forest network. We have learned that mother trees recognize and talk with their kin, shaping future generations. In addition, injured trees pass their legacies on to their neighbors, affecting gene regulation, defense chemistry, and resilience in the forest community. These discoveries have transformed our understanding of trees from competitive crusaders of the self to members of a connected, relating, communicating system. Ours is not the only lab making these discoveries—there is a burst of careful scientific research occurring worldwide that is uncovering all manner of ways that trees communicate with each other above and below ground.

Peter highlights these ground-breaking discoveries in his engaging narrative *The Hidden Life of Trees*. He describes the peculiar traits of these gentle, sessile creatures—the braiding of roots, shyness of crowns, wrinkling of tree skin, convergence of stem-rivers—in a manner that elicits an aha! moment with each chapter. His insights give new twists on our own observations, making us think more deeply about the inner workings of trees and forests.

DR. SUZANNE SIMARD
Professor of Forest Ecology
University of British Columbia, Vancouver
February 2016

ACKNOWLEDGMENTS

I VIEW THE fact that I can write so much about trees as a gift, because I learn something new every day as I research, think, observe, and draw conclusions from what I have discovered. My wife, Miriam, gave me this gift as she patiently took part in many conversations about what I was thinking, read the manuscript, and suggested countless improvements. Without my employer, the community of Hümmel, I would never have been able to protect the beautiful old forest that is my preserve, where I love to wander and which inspires me so much. I thank my German publisher, Ludwig, and publishers around the world for giving me the opportunity to make my thoughts available to a wide readership. And last but not least, I thank you, dear reader, for having explored some of the trees' secrets with me—only people who understand trees are capable of protecting them.

NOTES

1. S.W. Simard, D.A. Perry, M.D. Jones, D.D. Myrold, D.M. Durall, and R. Molina, "Net Transfer of Carbon between Tree Species with Shared Ectomycorrhizal Fungi," *Nature* 388 (1997): 579–82.

2. E.C. Fraser, V.J. Lieffers, and S.M. Landhäusser, "Carbohydrate Transfer through Root Grafts to Support Shaded Trees," *Tree Physiology* 26 (2006): 1019–23.

3. Massimo Maffei, quoted in M. Anhäuser, "The Silent Scream of the Lima Bean," *MaxPlanckResearch* 4 (2007): 65, www.mpg.de/942876/W001_Biology-Medicine_060_065.pdf, accessed February 16, 2016.

4. Ibid., 64.

5. M. Anhäuser, "The Silent Scream of the Lima Bean," *MaxPlanckResearch* 4 (2007): 62.

6. Y.Y. Song, S.W. Simard, A.Carroll, W.W. Mohn, and R.S. Zheng, "Defoliation of Interior Douglas-Fir Elicits Carbon Transfer and Defense Signalling to Ponderosa Pine Neighbors through Ectomycorrhizal Networks," *Nature, Scientific Reports* 5 (2015): art. 8495; and K.J. Beiler, D.M. Durall, S.W. Simard, S.A. Maxwell, and A.M. Kretzer, "Mapping the Wood-Wide Web: Mycorrhizal Networks Link Multiple Douglas-Fir Cohorts," *New Phytologist*, 185 (2010): 543–53.

7. Susanne Billig and Petra Geist, "Die Intelligenz der Pflanzen" ("The intelligence of plants"), *Deutschlandradio Kultur*, July 18, 2010, www.deutschlandradiokultur.de/die-intelligenz-der-pflanzen.1067.de.html?dram:article_id=175633, accessed December 12, 2014.

8. Tyroler Glückspilze, "Was sollte ich über die Anwendung von Mykorrhiza-Produkten wissen?" ("What should I know about using mycorrhizal products?"), www.gluckspilze.com/faq, accessed October 14, 2014.

9. S.W. Simard, D.A. Perry, M.D. Jones, D.D. Myrold, D.M. Durall, and R. Molina, "Net Transfer of Carbon between Tree Species with Shared Ectomycorrhizal Fungi," *Nature* 388 (1997): 579–82.

10. Ibid.

11. D.A. Perry, "A Moveable Feast: The Evolution of Resource Sharing in Plant–Fungus Communities," *Trends in Ecology & Evolution* 13 (1998): 432–34, and D.M. Wilkinson, "The Evolutionary Ecology of Mycorrhizal Networks," *Oikos* 82 (1998): 407–10.

12. N. Lymn, "Commercial Corn Varieties Lose Ability to Communicate with Their Own Defenders," Ecological Society of America, October 27, 2011, www.esa.org/esablog/research/commercial-corn-varieties-lose-ability-to-communicate-with-their-own-defenders, accessed January 26, 2016.

13. Monica Gagliano, et al., "Toward Understanding Plant Bioacoustics," *Trends in Plant Science* 17(6) (June 2012): 323–25.

14. Unpublished research from RWTH Aachen.

15. Knut Sturm, district forester, Lübeck, personal communication, 2015.

16. S. Dötterl, U. Glück, A. Jürgens, J. Woodring, and G. Aas, "Floral Reward, Advertisement and Attractiveness to Honey Bees in Dioecious *Salix caprea*," *PLoS One* 9(3) (2014): e93421.

17. Maurice E. Dermitt, Jr., "Poplar Hybrids," in Russell M. Burns and Barbara H. Honkala, tech. coords., *Silvics of North America*, vol. 2, *Hardwoods*, USDA Forest Service, 1990, www.na.fs.fed.us/pubs/silvics_manual/volume_2/populus/populus.htm, accessed January 30, 2016.

18. S.W. Simard, video, "Mother Tree," in Jane Engelsiepen, "'Mother Trees' Use Fungal Communication Systems to Preserve Forests," Ecology Global Network, October 8, 2012, www.ecology.com/2012/10/08/trees-communicate, accessed January 26, 2016, and S.W. Simard, K.J.

Beiler, M.A. Bingham, J.R. Deslippe, L.J. Philip, and F.P. Teste, "Mycorrhizal Networks: Mechanisms, Ecology and Modelling," *Fungal Biology Reviews* 26 (2012): 39–60.

19. Kenneth R.James, Nicholas Haritos, and Peter K. Ades, "Mechanical Stability of Trees under Dynamic Loads," *American Journal of Botany* 93(10) (October 2006): 1522–30, fig. 9, www.amjbot.org/content/93/10/1522.full, accessed January 30, 2016.

20. Max Planck Institute for Dynamics and Self-Organization, "Unterscheiden sich Laubbäume in ihrer Anpassung an Trockenheit? Wie viel Wasser brauchen Laubbäume?" ("Are there differences in how deciduous trees adapt to drought? How much water do deciduous trees need?"), www.ds.mpg.de/139253/05, accessed December 9, 2014.

21. University of Western Australia, "Move Over Elephants—Plants Have Memories Too," *University News*, January 15, 2014, www.news.uwa.edu.au/201401156399/research/move-over-elephants-mimosas-have-memories-too, accessed October 8, 2014.

22. Swiss Federal Institute for Forest, Snow, and Landscape Research WSL, "Rendering Ecophysiological Processes Audible," www.wsl.ch/fe/walddynamik/projekte/trees/index_EN, accessed January 26, 2016.

23. Swiss Federal Institute for Forest, Snow, and Landscape Research WSL, "Grösster Pilz der Schweiz" ("Largest fungus in Switzerland"), www.wsl.ch/medien/presse/pm_040924_DE, May 27, 2010, accessed December 18, 2014.

24. A. Casselman, "Strange but True: The Largest Organism on Earth Is a Fungus," *Scientific American*, October 4, 2007, www.scientificamerican.com/article/strange-but-true-largest-organism-is-fungus, accessed January 26, 2016.

25. U. Nehls, "Sugar Uptake and Channeling into Trehalose Metabolism in Poplar Ectomycorrhizae," dissertation, April 4, 2011, University of Tübingen.

26. SCINEXX, "Forscher belauschen Gespräche zwischen Pilz und Baum" ("Researchers eavesdrop on conversations between fungi and trees"),

January 23, 2008, www.scinexx.de/wissen-aktuell-7702-2008-01-23.html, accessed October 13, 2014.

27. S.W. Simard, D.A. Perry, M.D. Jones, D.D. Myrold, D.M. Durall, and R. Molina, "Net Transfer of Carbon between Tree Species with Shared Ectomycorrhizal Fungi." *Nature* 388 (1997): 579–82.

28. J. Fraser, "Root Fungi Can Turn Pine Trees into Carnivores—Or at Least Accomplices," *Scientific American,* May 12, 2015, http://blogs.scientificamerican.com/artful-amoeba/root-fungi-can-turn-pine-trees-into-carnivores-8212-or-at-least-accomplices, accessed January 26, 2016.

29. "Wassertransport in Gefäßpflanzen" ("Water transport in vascular plants"), www.chemgapedia.de/vsengine/vlu/vsc/de/ch/8/bc/vlu/transport/wassertransp.vlu/Page/vsc/de/ch/8/bc/transport/wassertransp3.vscml.html, accessed December 9, 2014.

30. K. Steppe, et al., "Low-Decibel Ultrasonic Acoustic Emissions are Temperature-Induced and Probably Have No Biotic Origin," *New Phytologist* 183 (2009): 928–31.

31. "Haut—Das Superorgan" ("Skin, the super-organ"), June 17, 2014, www.br-online.de/kinder/fragen-verstehen/wissen/2005/01193/, accessed March 18, 2015.

32. Zoë Lindo and Jonathan A. Whiteley, "Old Trees Contribute Bio-Available Nitrogen through Canopy Bryophytes," *Plant and Soil* (May 2011): 141–48.

33. J. Owen, "Oldest Living Tree Found in Sweden," *National Geographic News,* April 14, 2008, news.nationalgeographic.com/news/2008/04/080414-oldest-tree.html, accessed January 26, 2016.

34. František Baluška, et al., "Neurobiological View of Plants and Their Body Plan," in Baluška, Mancuso, Volkmann, eds., *Communication in Plants* (New York: Springer, 2007).

35. J. Copley, "Just How Little Do We Know about the Ocean Floor?" *Scientific American,* October 9, 2014, www.scientificamerican.com/article/

just-how-little-do-we-know-about-the-ocean-floor, accessed January 26, 2016.

36. "Faktensammlung: Bodendegradation" ("Summary: Soil degradation"), www.desertifikation.de/faktensammlung/fakten_degradation, accessed November 30, 2014.

37. Klara Krämer, thesis defense, RWTH Aachen, November 26, 2014.

38. A. Fichtner, et al., "Effects of Anthropogenic Disturbances on Soil Microbial Communities in Oak Forests Persist for More than 100 Years," *Soil Biology and Biochemistry* 70 (March 2014): 79–87.

39. National Fish and Wildlife Foundation press release, "NFWF Announces $4.6 Million in Funding for Restoration of Longleaf Pine Forest and Ecosystem across the Southeast," June 30, 2015, http://www.nfwf.org/whoweare/mediacenter/pr/Pages/longleaf-pr-15-0630.aspx, accessed February 12, 2016.

40. E.-Detlef Schultz, coord., John Gash, Annette Freibauer, Sebastian Luyssaert, Philippe Ciaia, eds., *CarboEurope-IP, An Assessment of the European Terrestrial Carbon Balance* (Jena: carboeurope.org, 2009), ftp://ftp.bgc-jena.mpg.de/pub/outgoing/athuille/CE_booklet_final_packed/CE_booklet_Stand_02-03-09_screen.pdf, accessed January 30, 2016.

41. Markus Johann Mühlbauer, seminar, "Wetter und Klima" ("Weather and climate"), in "Klimageschichte" ("Climate history"), WS 2012/13, 10, Regensburg University.

42. A. Mihatsch, "Neue Studie: Bäume sind die besten Kohlendioxidspeicher" ("Trees are the best carbon dioxide storage units"), press release, 008/2004, Leipzig University, January 1, 2014. See also, Becky Oskin, "Old Trees Grow Faster than Young Ones, New Study Shows," Huffpost Science, January 16, 2014, http://www.huffingtonpost.com/2014/01/16/big-trees-grow-faster-young_n_4609096.html, accessed February 10, 2016, and N.L. Stephenson, et al., "Rate of Tree Carbon Accumulation Increases Continuously with Tree Size," *Nature* 507 (March 6, 2014): 90–93, http://www.nature.com/nature/journal/v507/n7490/full/nature12914.html, accessed February 10, 2016.

43. L. Zimmermann, et al., "Wasserverbrauch von Wäldern" ("Water use in forests"), in *Wald und Wasser (Woods and Water), LWF-aktuell-66* (Freising: Bayerischen Landesanstalt für Walt und Forstwirtschaft [Bavarian State Institute of Forestry], 2008), 16, www.lwf.bayern.de/boden-klima/bodeninventur/012063/index.php, accessed February 16, 2016.

44. A.M. Makarieva and V.G. Gorshkov, "Biotic Pump of Atmospheric Moisture as Driver of the Hydrological Cycle on Land," *Hydrology and Earth System Sciences*, 11(2) (2007): 1013–33, www.bioticregulation.ru/common/pdf/07e01s-hess_mg_.pdf, accessed February 16, 2016.

45. D. Adam, "Chemical Released by Trees Can Help Cool Planet, Scientists Find," *Guardian*, October 31, 2008, www.theguardian.com/environment/2008/oct/31/forests-climatechange, accessed December 30, 2014.

46. T. Zhao, K. Axelsson, P. Krokene, and A.K. Borg-Karlson, "Fungal Symbionts of the Spruce Bark Beetle Synthesize the Beetle Aggregation Pheromone 2-Methyl-3-buten-2-ol," *Journal of Chemical Ecology* 41(9) (September 2015): 848–52, http://link.springer.com/article/10.1007%2Fs10886-015-0617-3, accessed January 30, 2016.

47. G. Möller, "Grosshöhlen als Zentren der Biodiversität" ("Large tree cavities as centers of biodiversity") (2006), www.biotopholz.de/media/download_gallery/Grosshoehlen_-_Biodiversitaet.pdf, accessed December 27, 2015.

48. Martin Gossner, et al., "Wie viele Arten leben auf der älteste Tanne des Bayerischen Walds?" ("How many species live on the oldest pine in the Bavarian Forest?"), *AFZ-Der Wald* 4 (2009): 164–65.

49. G. Möller, "Grosshöhlen als Zentren der Biodiversität" ("Large tree cavities as centers of biodiversity") (2006), www.biotopholz.de/media/download_gallery/Grosshoehlen_-_Biodiversitaet.pdf, accessed December 27, 2015.

50. Swiss Federal Institute for Forest, Snow, and Landscape Research, "Totholz und alte Bäume" ("Dead wood and old trees"), www.totholz.ch, accessed December 12, 2015.

51. Marco Archetti, "The Origin of Autumn Colours by Coevolution," *Journal of Theoretical Biology* 205(4) (August 21, 2000): 625–30, www.ncbi.nlm.nih.gov/pubmed/10931756, accessed January 30, 2016.

52. "Die Anatomie des Laubblattes" ("The anatomy of a deciduous leaf"), http://tecfaetu.unige.ch/perso/staf/notari/arbeitsbl_liestal/botanik/laubblatt_anatomie_i.pdf, accessed January 30, 2016.

53. H. Claessens, "L'aulne glutineux (*Alnus glutinosa*): une essence forestière oubliée" ("The common alder [*Alnus glutinosa*]: a forgotten forest fundamental"), *Silva belgica* 97 (1990): 25–33.

54. J. Laube, et al., "Chilling Outweighs Photoperiod in Preventing Precocious Spring Development," *Global Change Biology* 20(1): 170–82.

55. *National Geographic Germany*, "Woher wissen die Pflanzen wann es Fruehling wird?" ("How do the flowers know it's spring?"), March 9, 2012, www.nationalgeographic.de/aktuelles/woher-wissen-die-pflanzen-wann-es-fruehling-wird, accessed November 24, 2015.

56. Christoph Richter, "Phytonzidforschung—ein Beitrag zur Ressourcenfrage" ("Phytoncide research—a contribution to resource questions"), *Hercynia N.F.* 24(1) (1987): 95–106.

57. P. Cherubini, et al., "Tree-Life History Prior to Death: Two Fungal Root Pathogens Affect Tree-Ring Growth Differently," *Journal of Ecology* 90 (2002): 839–50.

58. T. Stützel, et al., *Wurzeleinwuchs in Abwasserleitungen und Kanäle (Tree roots growing in sewer pipes and tunnels)* (Gelsenkirchen: Ministerium für Umwelt, Naturschutz, Landwirtschaft und Verbraucherschutz des Landes Nordrhein-Westfalen [North Rhine Westphalia Ministry for Environment, Nature, Agriculture, and Consumer Protection], July 2004), 31–35, http://www.ikt.de/website/down/f0108kurzbericht.pdf, accessed February 16, 2016.

59. T. Sobcsky, "Der Eichenprozessionsspinner in Deutschland" ("The oak processionary in Germany"), *BfN-Skripten* 365 (Bonn-Bad Godesberg: Bundesamt für Naturschutz [Federal Agency for Nature Conservation],

May 2014), www.bfn.de/fileadmin/MDB/documents/service/Skript_365.pdf, accessed February 16, 2016.

60. Sandra Ebeling, et al., "From a Traditional Medicine Plant to a Rational Drug: Understanding the Clinically Proven Wound Healing Efficacy of Birch Bark Extract," *PLoS One* 9(1) (January 22, 2014): e86147, www.ncbi.nlm.nih.gov/pubmed/24465925, accessed February 16, 2016.

61. Michael G. Grant, "The Trembling Giant," *Discover*, October 1993, http://discovermagazine.com/1993/oct/thetremblinggian285, accessed January 25, 2016.

62. G. Meister, *Die Tanne (The Fir)* (Bonn: Schutzgemeinschaft Deutscher Wald (SDW) [The German Association for the Protection of Forests and Woodlands], nd), www.sdw.de/cms/upload/pdf/Tanne_Faltblatt.pdf, accessed February 15, 2016.

63. Reiner Finkeldey and Hans H. Hattemer, "Genetische Variation in Wälder—wo stehen wir?" ("Genetic variation in forests—where are we?"), *Forstarchiv* 81 (July 2010): 123–28.

64. James K. Agee, *Forest Fire Ecology of Pacific Northwest Forests* (Washington D.C.: Island Press, 1993).

65. Frank Harmuth, et al., "Der sächsische Wald im Dienst der Allgemeinheit" ("The woods of Saxony in the service of the general public") (Dresden: Staatsbetrieb Sachsenforst [State Forestry Commission of Saxony], 2003), 33, www.smul.sachsen.de/sbs/download/Der_saechsische_Wald.pdf, accessed February 15, 2016.

66. A. von Haller, *Lebenswichtig aber unerkannt (Necessary for life yet unknown)* (Langenburg: Boden und Gesundheit, 1980).

67. Jee-Yon Lee and Duk-Chul Lee, "Cardio and Pulmonary Benefits of Forest Walking versus City Walking in Elderly Women: A Randomized, Controlled, Open-Label Trial," *European Journal of Integrative Medicine* 6 (2014): 5–11.

68. Wilhelmshaven Botanic Garden, "Wassertransport" ("Water transport"), www.wilhelmshaven.de/botanischergarten/infoblaetter/wassertransport.pdf, accessed November 21, 2014.

69. S. Boch, et al., "High Plant Species Richness Indicates Management-Related Disturbances Rather Than the Conservation of Forests," *Basic and Applied Ecology* 14 (2013): 496–505.

70. Mark Hume, "Preserving the Great Bear Rainforest Doesn't Really Save the Bears," *Globe and Mail*, February 8, 2016, www.theglobeandmail.com/news/british-columbia/preserving-the-great-bear-rainforest-doesnt-really-save-the-bears/article28662082/?cmpid=rss1, and Julie Gordon, "Historic Deal Protects Canada's Pacific Forest 'Jewel,'" *Christian Science Monitor*, February 11, 2016, www.csmonitor.com/World/Making-a-difference/Change-Agent/2016/0211/Historic-deal-protects-Canada-s-Pacific-forest-jewel, accessed February 12, 2016.

71. Osprey Orielle Lake, Women's Earth and Climate Action Network International, "Recognizing the Rights of Nature and the Living Forest," *Ecowatch*, February 6, 2016, http://ecowatch.com/2016/02/02/rights-of-nature-living-forest/, accessed February 17, 2016.

72. Bavarian National Park, "Nationalpark mit Wildwuchs" ("National park with wilderness"), www.br.de/themen/wissen/nationalpark-bayerischer-wald104.html, accessed November 9, 2015.

73. Süddeustche Zeitung, "Die Welt aus Sicht einer Fliege" ("A fly's eye view of the world"), May 19, 2010, www.sueddeutsche.de/panorama/forschung-die-welt-aus-sicht-einer-fliege-1.908384, accessed January 21, 2016.

74. Waldfreunde Königsdorf (Friends of the Königsdorf Forest), www.waldfreunde-koenigsdorf.de, accessed December 7, 2014.

75. Healing Harvest Forest Foundation, www.healingharvestforestfoundation.org, accessed February 15, 2016.

76. J. Robbins, "Why Trees Matter," *New York Times*, April 11, 2012, www.nytimes.com/2012/04/12/opinion/why-trees-matter.html?_r=1&, accessed December 30, 2014.

INDEX

acacia trees, 7
acorns, 19–20, 27, 28, 113–14, 150, 187–88, 190
Adirondack and Catskill parks, 233–34
age, indications of: crown, 65; mossy growth, 64–65; process of aging, 65–66; and root system, 81–82; wrinkles in bark, 62, 63
agriculture, modern, 11–12
air, forest, 221–22, 223–24, 225
alders, 78, 111, 143–44
algae, 95, 225
Alps, 103, 189
Amazonian rain forest, 107
ancient woodlands designation, 233
animals: distinction from plants, 83–84; rights of, 241–42. *See also* herbivores; insects; pests
annosus root rot, 158–59
anthocyanin, 229–30
ants: and aphids, 116, 119; habitation in wood, 129; for pest control, 118–19; red wood, 219–20
aphids, 115, 116, 119
Arctic shrubby birch, 80
ash dieback fungus, 216–17
ash trees, 144, 187, 216–17
Asian long-horned beetle, 216
aspen, quaking, 181, 183–84. *See also* pioneer tree species
Australia, 233

balanced systems, 93
bald cypress, 144–45
balsam, Himalayan, 218
Baluška, František, 83
bark: birch, 182; buds in, 149; and deer, 123–24; diseases of, 64; function of, 61; fungi entry through, 66; moisture retention in rough, 167; oak, 72; and pests, 115, 116; shedding of, 61–63; wrinkles in, 62, 63–64
bark beetles, 54, 119–20, 132, 157, 236–37
bats, 128
Bavarian National Forest, 237
bears: grizzly, 136; spirit, 234
beaver, 111
beech: bark of, 62, 63; climatic limitations for, 193–94; community needed for, 1–2, 15–18;

competitive nature of, 74, 76, 193–94; copper, 229–30; defense mechanisms against pests, 7–8; and Douglas firs, 214; and drought conditions, 77–78; estimating age of, 31–32, 63; growth strategy of, 33–34, 190; and humans, 190–91; leaf growth timing, 148; lifespan of, 155; and lightning, 205–6; microclimates created by, 99–100, 194; migration of, 189–90, 191; and moss, 168; and oaks, 69–70; pests for, 26–27, 115, 117; and pioneer species, 184–85; pruning of, 173; and rain, 102; reproduction by, 19–20, 25–26, 27–28, 29, 113–14, 187–88, 190; rest needed by, 142, 226; severely damaged, 71; small, 80; and water, 43, 57, 193; in wet conditions, 78, 111; winter preparations by, 144. *See also* deciduous trees
beech leaf-mining weevil, 26–27
beechnuts, 19–20, 27, 28, 29, 69, 113–14, 150, 187–88, 190
bees, 20–21, 23–24, 116
beetle mites, 88, 90
beetles: Asian long-horned, 216; bark, 54, 119–20, 132, 157, 236–37; black-headed cardinal, 55; blood-necked click, 129; habitation in wood, 129; hermit, 129–30; stag, 133; woodboring, 54–55, 70
betulin, 182
bicolored deceiver *(Laccaria bicolor)*, 54
biodiversity: failure to notice, 231; importance of, 53, 130; loss of, 232; in trees, 131–32
birch: Arctic shrubby, 80; bark of, 62, 182; and ice, 141; paper, 247–49; seeds of, 187; silver, 181, 182, 183, 185; in wet conditions, 78
bird cherry tree, 22–23, 28–29, 73, 118, 137
birds: chaffinches, 112; and conifers, 21, 192; dispersal of seeds and organisms by, 28, 90–91, 217; fieldfare, 217; in forests, 231; habitations in trees, 127; jays, 69, 113–14, 150, 187, 190, 192; nutcracker, 192; nuthatch, 127; red crossbills, 21; sapsuckers, 114; woodpecker, 54, 114, 125–26
black cherry, 213
black-headed cardinal beetle, 55
black poplar, 215
blackthorn, 181
blood-necked click beetle, 129
blood pressure, 223
blue skies, 227–28
boars, 19–20, 27, 72, 191
bracket fungus, 133–34
Brazil, 107
breathing, 224–25
British Columbia, 234
bumblebee hoverfly, 132
butterflies, 231

Caledonian Forest, 92
cambium, 45, 54, 119, 158, 159
capillary action, 56–57, 58
carbohydrates, 51, 114, 224
carbon 14 dating, 81
carbon dioxide, 93–94, 95–96, 224

caterpillars, 117–18, 177–78
Central Europe: forests in, 64, 234–35, 236
chaffinches, 112
character, tree, 152–53, 154
cherry trees: bird cherry, 22–23, 28–29, 73, 118, 137; black, 213; fall leaves of, 144; wild, 137
chestnut trees, 12, 187–88
chlorophyll, 1–2, 138, 228, 229
climate: abrupt changes in, 196–97; behavioral adaptations for, 197–98; genetic adaptations for, 198–99; microclimates, 99–100, 101, 107, 194; and tree migration, 188–90, 194. *See also* climate change; weather-related damage
climate change: forests as tool against, 97, 98, 107; and greenhouse gases, 96; and permafrost, 40; temperature rises from, 153, 188; worse case scenario, 196. *See also* climate
coal, 94, 95
color, 227, 228
commercial forests, *see* managed forests
communication: and brain in root system, 82–83; loss of, 11–12; via electrical signals, 8, 10, 12, 83; via root and fungal systems, 10–11, 51; via scent in humans, 6–7; via scent in trees, 7–9, 12; via sound, 12–13, 48
community, *see* friendship
conifers: adaptation to additional light, 46; aging in, 65; air filtration by, 156, 222; blood pressure under, 223; and ice, 141; ideal shape for, 37, 41; microclimates created by, 107; outside of natural environment, 219–20, 222; phytoncides from, 156; and process conservation, 236–37; and rain, 103; reproduction by, 19, 21–22, 187; sickness in, 157–58; and streams, 109–10; terpenes from, 107; water transport vessels in, 57; winter preparation by, 138, 144–45. *See also* fir; pines; spruce
conifer sawflies, 118
conservation: economics of, 91–92; examples of, 233–34; failure of human attempts, 211; interference in regeneration, 237; and misconceptions about forest appearances, 238–39; and open clearings, 232; process of forest regeneration, 235–38; public demand for, 243–44; and safety in forests, 239–40
copper beech, 229–30
coppicing, 80–81
coral, 95–96
cork oaks, 207
counting, ability to, 148
crowns: and aging, 65; on beeches, 69; on conifers, 102; in heavy rain, 202; on oaks, 71; pruning of, 173–74; shade from, 32; in storms, 38, 140; wetland habitats in, 132
cypress, bald, 144–45

damage, *see* diseases; injuries; weather-related damage
dawn redwood, 144–45
dead wood, 130, 133–35

death: end of life, 66–67; from herbivores, 50; from lack of rest, 142, 226; and reproduction, 27; strangulation from climbing plants, 36, 165; in urban areas, 175, 178; from winter storms, 139
deciduous trees: adaptation to additional light, 46; aging in, 65; and Asian long-horned beetle, 216; and bees, 20–21; evolution of, 139; growth strategies of, 41; ideal shape for, 37, 153–54, 203; and rain, 103; reproduction by, 19–20, 21, 25–26; sickness in, 157; and snow, 141; and streams, 109; and tornadoes, 202; water transport vessels in, 57; winter preparation by, 137–38, 144; and winter storms, 139–40. *See also* beech; oaks; willows
deer: and bark, 123–24; and silver firs, 193; and young trees, 35, 120–21. *See also* herbivores
defense mechanisms: in acacia trees, 7; in beech, 7–8; against climatic changes, 197–99; in community-oriented species, 182–83; in elms, 8–9; against fire, 207; against fungi, 153–54, 160; hidden reserves, 156; human sensing of, 222–23; in oaks, 7–8, 9, 10, 70–71; against pests, 7–9, 116, 118; phytoncides, 156; in pines, 8–9; of pioneer species, 181–82, 183, 185; in quaking aspen, 183–84; in silver birch, 182, 185; in spruce, 7–8, 119–20; in willows, 9
diseases, 64, 156–59. *See also* injuries; weather-related damage
dogs, 176–77
Douglas fir, 62, 145, 206, 211–12, 213–14, 247–49
dove, Eurasian collared, 217
drought, 27, 45, 77, 209
drunken forests, 41
dust, 167, 212, 221, 222
dwarf trees, 79–80

Ecuador, 234
elder trees, 144
electrical signals, 8, 10, 12, 83
elms, 8–9, 209–10
erosion, 87
etiquette, *see* shape, tree
Eurasian collared dove, 217
evolution, 195–96, 227

fever, 9
fieldfare, 217
fir: grand, 211–12; pests for, 115; and rain, 102; shedding of needles, 145; silver, 62, 65, 153, 192–93
fir, Douglas, 62, 145, 206, 211–12, 213–14, 247–49
fire, 206–9
fire salamander, 110
First Nations, 234
fish, 245
floods, 209–10
forest management, *see* conservation
forest preserves, 233–34
forestry industry, xiii. *See also* managed forests
forests: biodiversity in, 231–32; as carbon dioxide vacuum,

93–94; drunken, 41; human reactions to, 222–23; importance of, xi, 244–45; open areas in, 232; research on, 131–32, 249; as superorganisms, 3; as water pump, 106–7. *See also* conservation; managed forests; old-growth forests
fossil fuels, 94
freshwater snail, 107–8, 109
friendship: advantages of, 3–4; interconnection of roots, 2–3; levels of, 4–5; living stump example, 1–2; mutual support from, 15–16, 17–18, 249; and spacing of trees, 16–17
fruit flies, 242
fruit trees, 12, 148
fungi: introduction to, 50; and aphids, 116; and bark beetles, 119–20; and beetle mites, 88; defense against, 153–54, 160; host selection by, 52–53; lifespan of, 52; medical benefits from, 52; mycelium of, 50; partnership with, 2, 50–52, 54, 247, 248; and pinesap, 122; and pioneer species, 185; and pruned trees, 173–74; and resource redistribution, 16; resources taken from trees by, 51; and small cow wheat, 122; as threat, 66, 126, 157, 159–60; and toxins, 51–52; and tree communication, 10–11. *See also* fungi, types of
fungi, types of: annosus root rot, 158–59; ash dieback fungus, 216–17; bracket fungus, 133–34; honey fungus, 50, 121–22; *Laccaria bicolor* (bicolored deceiver), 54; oak milkcap, 50; red belt conk, 133–34. *See also* fungi

Gagliano, Monica, 12–13, 47–48
gall midges, 117
genetics: and climate changes, 198; crossbreeding between species, 214–15; modification of, 195
Germany, 65, 90, 235, 241
giant hogweed, 218
giant redwood, 169–73, 208
giraffes, 7
girdling, 17–18
Gossner, Martin, 131
grand fir, 211–12
grass, 123, 181
Great Bear Rainforest, 234
green color, 228–29
grizzly bear, 136
groundwater, 108
growth: adaptations to environment for, 74–75; and age, 97–98; of beech trees, 31–32; challenges for seedlings, 73–74; in commercial forests, 124; and competition with other species, 49, 53, 113–14; conventional wisdom on, 96–97; in fall, 142–43; and growth spurt stage, 34–35, 67; and herbivores, 35, 120–21; ideal conditions for, 74; impediments to, 35–36; learning from water deprivation, 44–45; learning to support itself, 45–46, 46–47; light deprivation for, 32–33; in middle story, 36; mother trees, 33–34, 64–65, 249; of pioneer species,

181; rest needed for, 43–44, 142, 226; sickness during, 156–57; slowness of, 33, 196; in spring, 143; of trunks, 163; and water, 43–44, 48, 193. *See also* reproduction; shape
Guatemala, 92

habitation, in trees: introduction to, 125; attempts to repair damage from, 128–29; for bats, 128; in dead wood, 130, 132–33, 134; for insects, 129–30; for nuthatch, 127; for owls, 128; research on, 131–32; and sound vibrations through wood, 127; wetland habitats in crowns, 132; for woodpeckers, 125–26
habitats, 219
harvesting, *see* logging
hazard beam, 203
headaches, 9
Healing Harvest Forest Foundation, 244
heartwood, 160
hemiparasites, 165–66
herbivores: and deciduous trees reproduction, 19–20, 27; and hunting, 191; and open spaces, 181; and plant death, 50. *See also* deer; pests
hermit beetles, 129–30
hibernation, 43–44, 142, 152, 225–26
Himalayan balsam, 218
hoarfrost, 204–5
hogweed, giant, 218
honey fungus, 50, 121–22
honeysuckle, 35–36, 165
hornbeam, 77–78, 187
housing, *see* habitation, in trees
hoverfly, bumblebee, 132
humans, 48, 97, 190–91, 207–9, 218, 222–23. *See also* conservation
humic acid, 110
Hümmel forest, xiv, 91–92, 217
hunting, 191

ice, 141, 210
ice ages, 188–89
Indigenous peoples, 234
injuries: from bark being eaten, 123–24; bark diseases, 64; defense against, 160–61; from falling trees, 159; and fungi, 66, 126, 157, 159–60; from lower branches, 154; reopening of old, 161; from salt, 177; to trunk, 159; from urine, 176–77; from use as rubbing posts, 122–23; from woodpeckers, 125–26. *See also* pests; sickness; weather-related damage
insects, 12, 61, 242. *See also* ants; beetles; pests
introduced species: arrival of, 211–12; establishment of, 217–18; and genetic crossbreeding, 214–15; inevitability of, 217; initial benefits for, 212; and native habitats, 219; uncertain outcomes for, 212–13, 215
ivy, 164–65

jackpine, 208–9
Japanese knotweed, 218
Japanese larch, 211–12, 214–15
jays, 69, 113–14, 150, 187, 190, 192

Kichwa people, Ecuador, 234
knotweed, Japanese, 218
Königsdorf forest, 243, 245

Laccaria bicolor (bicolored deceiver), 54
ladybugs, 116
Lametta effect, 158
language, 6. *See also* communication
larch: Japanese, 211–12, 214–15; shedding of needles by, 144–45
learning, by trees, 47–48. *See also* growth
leaves: anthocyanin in, 229–30; green color of, 228; growth of in spring, 147–49; and plankton, 245; winter loss of, 138, 139, 140–41, 142, 144
lichen, 168
light, *see* sunlight
lightning, 205–6, 207
Lindo, Zoë, 64
liverworts, 163–64
logging, 5, 14, 80–81, 94, 97, 243–44

Maffay, Peter, 132–33
Maffei, Massimo, 3
Makarieva, Anastassia, 106
managed forests: appearance of, 239; and bark beetles, 236–37; clearings in, 232; and fire, 207; growth in, 124; harvest rate in, 46; purpose of for industry, xiii; qualification as forests, 235; and red wood ants, 220; root networks in, 5; safety in, 239–40; spacing in, 14, 248; and storms, 201
maples: red in leaves of, 230; seeds of, 187; sugar, 58
mast years, 20
Matsunaga, Katsuhiko, 245
Maya Biosphere Reserve, 92
meadows, 209–10
medicinal properties: in betulin, 182; from phytoncides, 156; from salicylic acid, 9
memory, 149
mice, 187–88, 195
microclimates, 99–100, 101, 107, 194
migration: of beech, 189–90, 191; and climate, 188–90, 194; and habitat, 212; of silver fir, 192–93. *See also* introduced species
mimosas, 47–48
mistletoes, 165–66
mites, beetle, 88, 90
mortality, *see* death
mosquitoes, 156
moss, 64–65, 166–68
mother trees, 33–34, 64–65, 249
mountain ash, 80
mycelium, 50

native species, 218–19
nature preserves, 235
needles, 75–76, 144–45
new species, *see* introduced species
nitrogen, 54, 65, 144
North America: forest fires in, 208–9
nun moths, 117–18
nurse logs, 135
nutcracker, 192
nuthatch, 127

oak milkcap, 50
oak processionary, 177–78
oaks: bark of, 62, 72; and beech trees, 69–70; and black-headed cardinal beetle, 55; blood pressure under, 223; cork, 207; defense mechanisms, 7–8, 9, 10, 70–71; distress signal of, 68, 70; fall leaves of, 144; and floods, 209–10; and fungi, 50, 52–53; healthy growth of, 68–69; and ivy, 165; lifespan of, 155; and lightning, 205–6; pests for, 54–55, 70, 115, 117–18; regeneration of forests, 91; reproduction by, 19–20, 25–26, 27–28, 113–14, 187–88; resiliency of, 70–72; rest needed by, 142, 226; triad of near Hümmel, 151–53; in urban areas, 178; and woodpeckers, 54. *See also* deciduous trees
old-growth forests: designation as, 233; growing conditions in, 170; lack of in Central Europe, 64; misconceptions about appearance, 238–39; regeneration of, 89–90, 91–92, 235. *See also* conservation
open areas, in forests, 232
oribatid (beetle) mites, 88, 90
osmosis, 57, 58
owls, 128
oxygen, 223–24, 225

paper birch, 247–49
parks, *see* urban trees
pests: benefits for other animals, 116; caterpillars, 117–18, 177–78; climbing plants, 35–36, 164–66; conifer sawflies, 118; deer, 35, 120–21, 123–24; defense against, 7–9, 116, 118, 119; parasitical plants, 122; sap sucking insects, 115–16, 117, 119; spreading of, 215–16, 216–17; targeting of trees by, 11; and tree reproduction, 26–27; in urban areas, 177–78; variety of, 115; woodpeckers, 114. *See also* ants; beetles; fungi, types of; habitation, in trees; insects
photosynthesis, 15–16, 16–17, 35, 183, 224
phytoncides, 156, 223
pine loopers, 117–18
pines: bark of, 62–63; defense mechanisms, 8–9; and forest fires, 208–9; growth of, 41; and ivy, 164–65; jackpines, 208–9; and *Laccaria bicolor* (bicolored deceiver), 54; ponderosa, 208; shedding of needles, 145; sickness in, 158; in wet conditions, 78. *See also* conifers
pinesap, 122
pioneer tree species: competition with other species, 184–85; death of, 185; defense mechanisms, 181–82, 183–84, 185; and fungi, 185; growth rates of, 181; ideal sites for, 180–81, 188; propagation by, 180. *See also* quaking aspen; silver birch
plane tree, 178–79
plankton, 245

planted forests, *see* managed forests
plants: definition of, 49–50; distinction from animals, 83–84
pollination, 21–24. *See also* reproduction
pollutants, 51–52, 221–22
ponderosa pine, 208
pools, 110–11
poplars: black, 215; crossbreeding among, 215; as pioneer species, 181, 185, 188; reproduction by, 30; seeds of, 186; in wet conditions, 209
process conservation, 235. *See also* conservation
procreation, *see* reproduction
propagation, *see* reproduction
pruning, 173–74
pussy willow, 181

quaking aspen, 181, 183–84. *See also* pioneer tree species

rain, 101–2, 103, 107, 111–12, 202–3
red belt conk, 133–34
red crossbills, 21
redwood: dawn, 144–45; giant, 169–73, 208
red wood ants, 219–20
reforestation, 222
reproduction: by conifers, 19, 21–22; crossbreeding, 214–15; by deciduous trees, 19–21, 25–26; energy levels during, 25–26; and insects, 26–27; interbreeding, 22–24; and mortality, 27; odds for successful, 29–30; by pioneer species, 180; pollination, 21, 22–24; pre-planning for, 19–20; seeding strategies, 27–29, 186–88; from stumps, 80–81; timing of, 150. *See also* growth
reserves, hidden, 156
rest, 43–44, 142, 152, 225–26
rights, for plants, 242–43, 244
rivers, 209–10
root systems: and age, 81; as brain, 82–83; communication through, 10–11; depth of, 174–75; and fungi, 50–51; interconnections between, 2–3, 158–59, 206, 248; poor environments for, 73–74; and pruning of crowns, 173–74; and water, 49; in wet conditions, 73, 78. *See also* friendship
rubbing posts, trees as, 122–23

safety, in forests, 239–40
salamanders, 109, 110
salicylic acid, 9
salt, 177
sapsuckers, 114
savannah, African, 7
sawflies, conifer, 118
scent, as language, 6–7, 7–9, 12
school, tree, 47–48. *See also* growth
Scotland, 92
seeds, 27–29, 186–87
shadows, green, 229
shape, tree: curved trunks, 38; environmental effects on, 39–41; forked trees, 38–39; hazard beams, 203; ideal,

37–38, 153–54, 203; stability as goal, 38. *See also* growth
sickness, 64, 156–59. *See also* injuries; weather-related damage
sight, 148, 231
silver birch, 181, 182, 183, 185. *See also* pioneer tree species
silver firs, 62, 65, 153, 192–93
Simard, Suzanne, 9–10, 11, 33, 53, 247–50
Sitka spruce, 64–65
skin, 60–61. *See also* bark
sleep, 43–44, 142, 152, 225–26
Slett, Marilyn, 234
small cow wheat, 122
snail, freshwater, 107–8, 109
snow, 141, 203–4
social security, *see* friendship
soil: beetle mites in, 88; and carbon storage, 94, 95; coal formation in, 95; creation of, 86–87; erosion of, 87; importance of, 85–86; lack of knowledge about, 85; organisms in, 87–88, 89–91, 225; regeneration of after disruptions, 91; weevils in, 88–89
sound, 127
spirit bear, 234
spring (season), 143, 148
springs, 108–9
springtails, 54, 90
spruce: and age, 81–82; aging, 65; and climatic changes, 197; defense mechanisms, 7–8, 119–20; growth strategies of, 75–76; habitat for, 75, 219; pests for, 115; pollination of, 22; shedding of needles, 145; sickness in, 158; Sitka, 64–65; and sunlight, 167; and water, 44, 45, 102–3, 193; in wet conditions, 78; and woodpeckers, 54. *See also* conifers
squirrels, 127, 150, 187–88
stability, 38, 45–46, 46–47, 107–8, 130
stag beetle, 133
storms, 139–40, 153, 176, 201–2, 232
strangulation, 36, 165
streams, 109–10, 209
street kids, *see* urban trees
sugar maples, 58
sunlight: awareness of, 148–49; competition for, 162–63; deprivation of for growth, 32–33; and early blooming plants, 163–64; and green color of leaves, 228; importance of, 162; and ivy, 164–65; and mistletoe, 165–66; and moss, 167–68
sweating, 101
Switzerland, 244

taste, sense of, 9
temperature changes, 149, 153
terpenes, 107, 119
thinning, 14–15, 17–18
timing: confused sense of, 149–50; for leaf growth in spring, 147–49; for reproduction, 150; and temperature changes, 149; and tree character, 152–53
Tokin, Boris, 156
tornadoes, 202
transpiration, 57, 58, 106
trees: as balanced system, 93; difficulties defining, 79–80,

81–82; as guarded warehouse, 114; lifespan of, 155; misunderstanding of, 230–31; rights for, 242–43, 244. *See also* conifers; conservation; death; deciduous trees; defense mechanisms; friendship; growth; pests; reproduction; root systems; shape; urban trees; *specific species*
trunks, 159, 163. *See also* shape, tree

United Kingdom, 233
United States of America, 92, 233–34
urban trees: along streets, 174–75; and Asian long-horned beetle, 216; other challenges facing, 176–77; in parks, 169–73; and pests, 177–78; and pipes, 175–76; planting of same species together, 178–79; poor stability of, 176; premature death of, 178; and pruning, 173–74
urine, 176–77

walking sticks, 36, 165
walnut trees, 156
water: and conifers, 107; floods, 209–10; groundwater, 108; ice, 141, 210; importance of, 107, 193; learning to ration, 43–44, 44–45; pools, 110–11; rain, 101–2, 103, 107, 111–12, 202–3; and root systems, 49; snow, 141, 203–4; sound vibrations from in trees, 48; springs, 108–9; streams, 109–10; sweating by trees, 101; transportation of, 56–59, 105–6; and tree microclimates, 100–101; in trees during winter, 137
weather-related damage: fire, 206–7, 208–9; floods, 209–10; heavy rain, 202–3; hoarfrost, 204–5; lightning, 205–6, 207; tornadoes, 202; wet snow, 203–4; winter storms, 201. *See also* climate
weevils, 88–89. *See also* beech leaf-mining weevil
wild cherries, 137
wild service tree, 137
willows: defense mechanisms, 9; as pioneer species, 188; pollination of, 23–24; pussy, 181; scents for attracting insects, 12; seeds of, 186; in wet conditions, 111, 209
wind, 21, 38, 76, 100, 139–40. *See also* storms
winter preparation, 136–38, 144–45
wolves, x–xi, 218–19
wood anemones, 163
woodboring beetles, 54–55, 70
woodpeckers, 54, 114, 125–26
wood wide web, 10–11, 249
woolly beech scale, 115

Yellowstone National Park, x–xi
yew, 76–77

THE INNER LIFE OF ANIMALS

Love, Grief, and Compassion
Surprising Observations of a Hidden World

A BEGUILING AND revelatory new book from the internationally bestselling author of *The Hidden Life of Trees* that opens up the animal kingdom like never before.

We humans tend to assume that we are the only living things able to experience feelings intensely and consciously. But have you ever wondered what's going on in an animal's head?

From the leafy forest floor to the inside of a bee hive, *The Inner Life of Animals* shows us microscopic levels of observation as well as forcing us to confront the big philosophical, ethical and scientific questions. We hear the stories of a grateful humpback whale, of a hedgehog who has nightmares, and of a magpie who commits adultery; we meet bees that plan for the future, pigs who learn their own names and crows that go tobogganing for fun. And at last we find out why wasps exist.